TRACTORS

TRACTORS

John Carroll and Garry Stuart

PB
PARKGATE
BOOKS

First published in 2000 by
PRC Publishing Ltd,
8-10 Blenheim Court Brewery Road London N79NY

This edition published in 2000 by
Parkgate Books
London House
Great Eastern Wharf
Parkgate Road
London
SW11 4NQ

British Library Cataloguing in Publication Data:
A catalogue record for this book is available from the
British Library.

ISBN 1 902616 65 0

Printed and bound in Hong Kong

Acknowledgments
The authors and photographers involved with this book are
grateful to the owners of all the tractors photographed,
Dwayne Mathies and to the organisers of vintage
agricultural events in both England and America including
those in the English Cotswolds and South Dakota as well as
the Florida Flywheelers, Pacific Coast Dream Machines and
the Masham Steam Engine and Fair Organ Rally.
John Carroll and Garry Stuart are especially grateful to
Michael Thorne of the Coldridge Collection, Crediton,
Devon, EX17 6TS, England (+44 (0) 1363 83418) for his
enormous help with this project despite its short deadlines.
Photographs by Garry Stuart, John Carroll, John Bolt and
Wayne Mitchelson. Archive photographs from Ian Allan
Collection, Massey Ferguson and John Carroll.

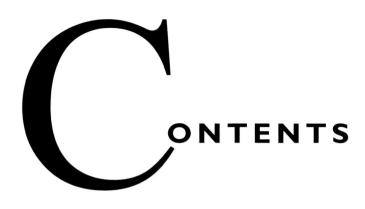

CONTENTS

TRACTORS

INTRODUCTION

The development of the tractor is inextricably linked with the development of machinery in general. Early developments in technology were applied in all fields of endeavour—steam engines, for example, were used to power factories, ships, trains, and vehicles—and it was only later that developments became considerably more specialised.

It is acknowledged that the era of mechanically propelled transport started in 1769 when Nicholas Cugnot built a three-wheeled vehicle with steam propulsion. This was the first machine specifically designed for haulage; it was capable of carrying four passengers at 4mph and ran in the streets of Paris. It can be considered as the first tractor, the first truck, the first military vehicle (it was intended as a gun tractor), the first bus and so on. So it was for many other early machines.

Far Left: Vintage John Deere tractor at a Florida Flywheelers' event.

Left: Weathered McCormick-Deering logo suggests years of hard work.

A Welsh inventor, Oliver Evans who had emigrated to America and lived in Maryland, produced a elementary steam wagon in 1772. He looked into the possibilities of applying steam power to propel a vehicle. In 1787 he was granted the right to manufacture a steam wagon. His wagon never made it as far as production but he did build a steam-powered dredging machine that he rigged to drive from its place of manufacture to the River Schuylkill before been driven to Delaware. In 1788 a vehicle built along similar lines—the Fourness—was built in Great Britain. By 1831 the idea was proven and men such as Sir Charles Dance and Walter Hancock operated a number of steam coaches on regular routes. The latter's machines were capable of up to 20mph. The railway age boomed but the mechanisation of road transport and other machinery—such as that used in agriculture—was more difficult to arrange in practice and experimentation continued in both Europe and the USA.

It wasn't until the closing years of the 19th century that vehicles powered by the internal combustion engine started to make an appearance. Names like Benz, Daimler, De Dion, Panhard and Peugeot became prominent in Europe, and in

Below: A 1906 Russell at a vintage tractor event in California.

Above right: Old timers at a vintage tractor rally in South Dakota.

Below right: A vintage John Deere being used for a ploughing demonstration at the same event.

Above: Port Huron Threshing engine at the Pacific Coast Dream Machines show in Half Moon Bay, California.

Right: Vintage tractor ploughing. Note that numerous adjustments can be made to the plough to cope with different soil types and field conditions.

Britain Albion, Dennis, Humber, Napier, Sunbeam and Wolseley. The first British commercial vehicle that was both viable and practical was made by Thornycroft in 1896. This was the same year as the 'Red Flag Act' was repealed by an English parliament. This act had forced internal combustion engine vehicle operators to employ a man to walk in front of them with a flag warning of danger. The repeal of this repressive legislation paved the way for the development of both the

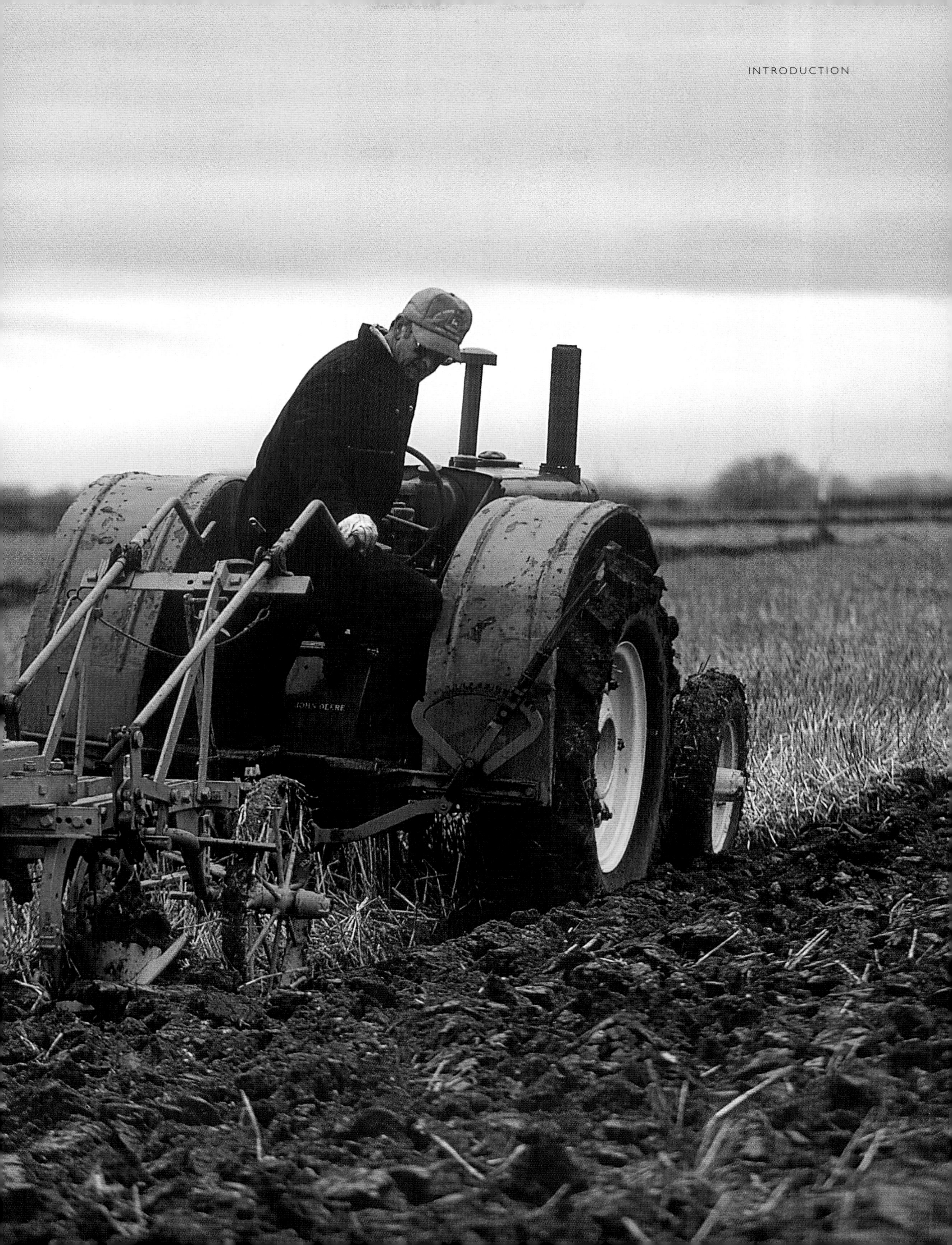
JOHN DEERE

Left: A John Deere seed drill behind a John Deere tricycle row-crop tractor planting four rows of seeds at once.

Below: A 1920s John Deere GP with sidebar mower being used in the construction of a haystack.

steam and internal combustion engine. A year earlier Richard F. Stewart of Pocantico Hills, New York produced a truck with a 2hp Daimler engine and internal gear drive for his first truck, and two years later he started producing trucks for sale. This was a steam-powered truck with a marine-type steam boiler and vertical engine.

Within a few years Thorneycroft had produced the world's first articulated truck. The British Army was quick to realise the potential of such machines

Above: Industrial Designer Henry Dreyfuss restyled the John Deere Models A and B to update their appearance and they stayed in production between 1935 and 1952.

Right: Vintage Tractors, including a tricycle row-crop Case, parade at the famous Masham, Yorkshire, England traction engine rally.

and by 1899 had purchased some for use in the Boer War in South Africa. they used steam traction engines for hauling artillery. Leyland vehicles appeared in 1896 under the auspices of the Lancashire Steam Motor Company, whose first vehicle was a van with an oil-fired boiler and a two-cylinder compound engine. The steam traction engine manufacturers began to build steam lorries and one of them, Foden, was to become the world's largest steam vehicle manufacturer.

In 1897 the Daimler Motor Co. of Coventry, England offered a petrol-engined commercial vehicle. It was designed by Panhard and powered by a Daimler internal combustion engine. From this vehicle they went on to build other similar machines. Across the Atlantic the first Mack truck was rolling out of a Brooklyn works. The company had been established by five brothers of German parentage who had formerly run a Brooklyn smithy. The smithy was gradually turned over to

the production of trucks, and their first is reputed to have travelled a million miles.

In 1904 the brothers built a charabanc which they named the Manhattan and by 1905 had been sufficiently successful to move their operation to Allenstown, Pennsylvania where they introduced the Model AC. This was the truck that earned them the 'Bulldog Mack' nickname. It was a four-cylinder petrol-engined truck based on a pressed-steel chassis frame. Transmission was by means of a three-speed gearbox through a jackshaft which had chain drive to rear wheels fitted with solid tires. The truck featured a cab—unusual at the time—of a bonneted design. The AC was supplied in significant numbers to the British Army in France where it earned the 'Bulldog' tag. Another landmark from Mack was the predecessor of the 'cab over engine' design, in that it was realised that if the driver sat over the engine it was possible to incorporate a longer load bed in a chassis of the

same length. The fledgling Mack company also developed the Junior model, a two-ton truck intended for lighter duties which in many ways was the forerunner of the delivery van while other companies developed more specialised machines for farm use.

From common beginnings, trucks, buses, cars, vans and, the subject of this book, tractors were derived and subsequently diverged. In a period of time that spans little more than a century, tractors have gone from being primitive and unreliable to utterly sophisticated and precise machines. The latest innovations include turbo-diesel engines, four-wheel drive transmissions, computer-controlled functions and operator comforts such as vibration-proof, air-conditioned cabs.

Left and Below: Preserving vintage and classic tractors is a worldwide phenomenon as these photos from Florida (**above left**), England (**below left**) and South Dakota (**below**) illustrate.

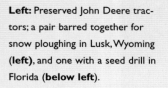

Left: Preserved John Deere tractors; a pair barred together for snow ploughing in Lusk, Wyoming **(left)**, and one with a seed drill in Florida **(below left)**.

The mechanisation of farming goes right back to events such as Jethro Tull's invention of the seed drill and Andrew Meikle's development of a machine that later became the mechanical thresher. The need for a mechanical source of power to drive early machines was clearly understood and, after experimentation, steam power proved the most suitable; soon machines such as Thomas Newcomen's steam pump were operational. The concept was refined by James Watt and Nicholas Cugnot. The advent of the railway locomotive again focussed attention on the possibilities of steam-powered machinery that was independent of rails. Gradually the steam traction engine became more refined and a practical proposition for farm use. However, in the main, the steam traction engine was reserved for providing mobile stationary power in driving threshing machines and similar.

The obvious exception to this was the pairs of ploughing engines that were employed for field cultivation. These relied on drawing a plough backwards and forwards between them by a winch system. This method endured in many areas until the 1930s. However, the size of farms in North America and South Africa meant that the need for further mechanisation still existed. Progress came in the shape of the internal combustion engine and associated developments of the time. From the workable internal combustion engine it was but a short step to a practical agricultural tractor.

Much of the early development concerned threshing, a labour-intensive process prior to mechanisation. John Charter built petrol engines in Stirling, Illinois and manufactured a tractor by fit-

Above right: A familiar farm scene; a well used and ageing Allis-Chalmers alongside a Fordson Dexta in an English farmyard barn. Both are still ready for work.

Below right: The combine harvester such as this can be considered as a specialised tractor. It has to cope with field conditions but engineered for one farming task, namely harvesting.

Below: The 20 series of John Deere tractors were announced in 1956. The range comprised of various models ranging from the 320 to the 820 with different horsepower ratings.

ting one of his engines to the chassis and wheels of a steam traction engine. The machine was put to work on a wheat farm in South Dakota in 1889. It was a success and Charter is known to have built several more machines to a similar specification.

By 1892 a number of other fledgling manufacturers were starting to produce tractors powered by internal combustion engines. John Froelich built a machine powered by a Van Duzen single-cylinder engine in Iowa that many consider as the first practical tractor. The engine of the machine was mounted on the Robinson chassis from a steam engine and Froelich devised a transmission system. He was experienced in the agricultural business and had worked as a threshing contractor so was aware of the requirements of mechanised harvesting. He bought a living wagon and a large Case thresher and transported his machinery by rail to Langford, South Dakota. It is reported that hundreds turned out to see the machines working when over a seven-week period he threshed

wheat full time. His machinery suffered no breakdowns and, as a result, many were convinced of the benefits of mechanisation. Froelich gained backing from a group of Iowa businessmen and they formed the Waterloo Gasoline Traction Engine Co. In 1893 the company built four more tractors of which only two were fully workable, and others in 1896 and 1897. The company later dropped the word 'traction' from its name in 1895 and concentrated on the manufacture of stationary engines. John Froelich's interest was primarily in tractors so he left the company at this time.

The J. I. Case Threshing Co. was formed in 1863 to build steam tractors. Their first tractor

Previous page: A 1950s Cockshutt Black Hawk 35 row-crop tractor. The American Cockshutt company was acquired by White in 1962.

Right: An ageing Farmall tractor in front of a windmill.

Below: There have been numerous specialist compact tractors produced over the years.

appeared in 1892 but it was not until the beginning of the second decade of the 20th century that Case built gasoline-fuelled machines. It was a similar story for numerous other companies; Case experimented with gasoline engines using one designed by William Paterson, and then went back to steam. Other tiny companies built prototypes but achieved little more until the Huber Co. of Marion, Ohio purchased the Van Duzen Engine Co. and built a batch of 30 tractors. A couple of other firms started around this time too, including the Otto Gas Engine Co. and the Kinnard-Haines concern from Minneapolis. Two other companies, Deering and McCormick, built self-propelled mowers at this time; the speed of mechanisation of American farming was increasing.

Massey-Harris was formed in 1891 in Toronto, Canada from the merger of two competing companies who manufactured farm implements. Daniel Massey had been making implements since 1847 while A. Harris, Son & Co. were competitors for

the same market. Similar developments were taking place in Europe; in 1873 Laverda was founded in Breganze, Italy and by the beginning of this century had become the leading Italian manufacturer of threshing machines. The eponymous Landini company—the oldest established tractor manufacturer in Italy—was founded by Giovanni Landini, a blacksmith who opened his own business in Fabbrico in Italy's Po Valley during 1884. Giovanni Landini began a mechanical engineering concern, which was successful and gradually he progressed from blacksmithing to fabricating farm machinery for local farms. He progressed further by producing winemaking machinery, then steam engines, internal combustion engines and crushing equipment. In 1870 Braud was founded in France, in St. Mars La Jaille, to manufacture threshing machines. The development of increasingly viable machines, the right economic conditions for its production and use and within a few years the tractor would be in business in the fields. A century later, it has never left them.

Opposite: This Bradley machine was manufactured for specialist farming applications as were the array of compact and garden tractors (**above and left**).

TRACTORS

1

THE EARLY YEARS

The economic conditions prevalent in farming on both sides of the Atlantic before World War I meant that North America was where the majority of tractor production took place. From early in the twentieth century tractor trials—to evaluate tractor performance and make realistic comparisons between the various models—were held. Following the success of the Canadian Winnipeg Trials of 1908, these continued

Left: Steam power as this Nicholls and Shepard, gradually gave way to kerosene, gasoline and diesel propulsion as the tractor was developed into a viable farm tool.

Opposite, Above: A preserved International Harvester Farmall Cub at a rally in South Dakota.

Opposite, Below: A John Deere Model A in England.

as a regular event until 1912, while in the United States trials were held in Nebraska. The Nebraska tractor tests became established as a way of determining the capabilities of tractors and preventing their manufacturers from claiming unlikely levels of performance. Starting in 1920 a series of tests on tractors were undertaken that examined horsepower, fuel consumption, and engine efficiency. There were also practical tests that gauged the tractor's abilities with implements on a drawbar. These tests were carried out at the Nebraska State University in Lincoln, Nebraska. The Nebraska State law decreed that manufacturers must print all or none of the tester's results ensuring that a manufacturer could not publish the praise and delete the criticism. These tests were

noted for their fairness and authority and led to their general acceptance far beyond the State of Nebraska.

The United States attempted to remain neutral in World War I, but sympathy for the Allied cause ultimately brought the United States into the war against the Germans. From the point of view of manufacturers there was the prospect of sales to the military of every type of motor vehicle. It was no secret that the US War Department had been experimenting with automobiles as a result of the 1914 war with Mexico that had led to the American occupation of Vera Cruz. America declared war on Germany and its allies on April 6, 1917. The Democrat President, Thomas Woodrow Wilson, made a plea to industry requesting all-out

production and as a result production reached an all time high. The tractor market was also changing. The realisation that smaller tractors were practical changed the emphasis of the industry and threatened some of the old established companies.

When the war ended in 1918, major socio-economic changes that would change the face of farming came in its wake; on top of this, there had been huge progress in the engineering and automotive industries. Many of the fledgling prewar vehicles had been transformed into practical and functional machines. Tyre technology, transmissions, engines and the like had all progressed considerably and tractor manufacturers would be able to further exploit these developments.

Prosperity followed the war and the number of tractor manufacturers around the world mushroomed. Many were tiny companies, with little chance of success, but the 1920s were the time when the tractor industry saw mass-produced machines gaining sales everywhere. Henry Ford's Fordson sold in vast numbers achieving some 75% of total tractor sales. Having fulfilled his order for the British military, Ford had started selling his proven tractor into the domestic American market. It was cheap to produce and sold cheaply, meaning that it was affordable to a greater number of farmers.

The 1920s saw much experimentation with four-wheel drive tractors as an alternative to crawler machines. Wizard, Topp-Stewart, Nelson and Fitch were amongst those who manufactured machines.

Left: This Nicholls and Shepard machine is equipped with a driven wheel for powering far machinery such as threshers, and steel lugged wheels to aid traction in fields.

This chapter charts the rise of tractor-manufacturing companies in the early years and the mergers and acquisitions that would lead to the domination of a few big companies. While many of the tiny manufacturers struggled with production in handfuls, the tractor market developed into a struggle between Fordson, International Harvester, Case and John Deere.

North America
ACME

One of a lengthy list of shortlived tractor manufacturers founded during the boom years that followed the end of World War I. The ACME tractor was advertised as being available in both wheeled and half-track form. Few were made.

Allis-Chalmers

Allis-Chalmers has roots that stretch back to Milwaukee's Reliance Works Flour Milling Co. which was founded in 1847. Allis-Chalmers was formed in 1901 by the merger of four other companies. The new company was reorganised in 1912 with Brigadier-General Otto H. Falk as president. It remained based in Milwaukee, Wisconsin and although it had no experience of steam traction built its first gasoline tractor, the tricycle-type Model 10-18, in 1914. This had a two-cylinder opposed engine that revved to 720rpm and, as its designation indicates, produced 10 drawbar horsepower and 18 belt horsepower. (All the model numbers of this style—12-25—indicate the output of 12hp at the drawbar and 25hp at the belt. This means that info to come info to come info to come info to come info to come info to come.) It was started on gasoline and once the engine warmed up it ran on cheaper kerosene. The single wheel was at the front while the driver sat over the rear axle.

Unlike some of its competitors, Allis-Chalmers did not have an established dealer network around the United States, so sales did not achieve their full potential. Between 1914 and 1921 the company manufactured and sold only around 2,700 10-18s. Some of these were sold through mail order catalogues and during World War I some export sales were achieved. The French imported American tractors and sold them with French sounding brand names, the Allis-Chalmers 10-18 was marketed as the Globe tractor.

The conventional Model 18-30 tractor was introduced in 1919, it was powered by a vertical in-line four-cylinder engine. Initially sales were limited, partly as a result of competition from the cheap Fordson, and slightly more than a thousand had been assembled by 1922, but over the course of the next seven years the total reached approximately 16,000. Also introduced in 1919 was the Model 6-12.

Allis-Chalmers acquired a few other companies when the end of the postwar boom led to numerous closures and mergers within the tractor industry. One of these acquisitions was the Monarch Tractor Co. of Springfield, Illinois which made crawler tractors. The company had started tractor production in 1917 and had reorganised twice by the time of the merger, when production of a range of six different sized crawlers was underway. The smallest of these was the Lightfoot 6-10 and the largest the Monarch 75, which weighed 11.5 tons (10,350kg). Allis-Chalmers continued the production of crawlers in Springfield.

In 1929 as many as 32 farm equipment making companies merged to form the United Tractor and Equipment Corporation which had its headquarters in Chicago, Illinois. Amongst the 32 was Allis-Chalmers, which was contracted to build a new tractor powered by a Continental engine and known as the United. The tractor was launched at an agricultural show in Wichita in the spring of 1929. The corporation did not stay in business long. Allis-Chalmers was fortunate enough to survive the collapse and continued to build the United tractor, albeit redesignated the Model U. The models U and E became the basis of the Allis-Chalmers range. Allis-Chalmers also introduced a bright orange colour scheme to attract new customers and differentiate its tractors from those of other makers. The colour chosen by Allis-Chalmers was called Persian Orange It was a simple ploy but one that no doubt worked as other manufacturers soon followed suit with bright coloured paintwork and stylised bonnets, radiator grilles and mudguards.

Allis-Chalmers acquired other companies during the 1930s including the Advance Rumely Thresher Co. It had its origins in the form of Meinrad Rumely—a German emigrant who ran a blacksmith's shop in Laporte, Indiana during the 1850s. The smithy expanded into a factory that produced machines and steam engines for agricultural use. In 1908 John Secor joined Rumely to develop an oil-fuelled engine. The company was primarily a manufacturer of threshers and made its first tractor in 1909 after Secor's nephew had perfected a carburettor for kerosene or paraffin fuel. A later result of this development work was the Model B 25-45 tractor which was superseded by the Model E of 1911.

The Model E was a 30-60 tractor. These figures were substantiated when the Model E was tested in the 1911 Winnipeg Agricultural Motor

Right: Case was a manufacturer of steam engines for agricultural use and its emblem of the eagle 'Old Abe' is clearly evident on the front of this machine.

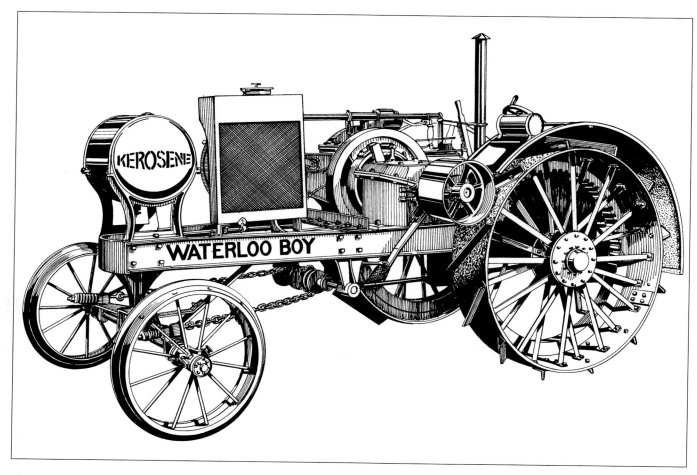

Competition. In 1920 the same tractor was the subject of Nebraska Tractor Test No. 8, when the drawbar figure was measured at almost 50hp and the belt at more than 75hp. The engine capable of producing this horsepower was a low revving horizontal twin-cylinder with a bore and stroke of 10in x 12in. The measured fuel consumption of this engine was high at almost 11 gallons of kerosene per hour. Notable features of the engine design included the special carburettor with water injection and induced air cooling by creating a draft through the rectangular tower on the front of the tractor.

The company was renamed as the Advance-Rumely Thresher Co. in 1915 during the production run of the Model E 30-60 Oil Pull, which remained in production until 1923 when it was superseded by the similarly designed 20-40 Model G. By 1931 Advance-Rumely had produced in

Above: A 1916 Waterloo Boy Model N tractor. John Deere later bought the Waterloo Boy company in order to move into tractor manufacture, so becoming a 'full line' agricultural implement maker.

Above right: Charles Hart and Charles Parr began developing gasoline tractors in the late 1800s in Wisconsin. This is a Hart Parr tractor from the early 1900s.

excess of 56,500 Oil Pulls in 14 configurations. Advance-Rumely entered the small tractor market in 1916 when it first advertised its all purpose 8-16 model. The operation of the machines was described in advertising material of the time as being, 'just like handling a horse gang'. The machine had only three wheels, a single steering rear wheel and two front wheels, of which one was driven and the other free-wheeled. It was powered by a four-cylinder engine that ran on kerosene. The machine was intended for drawbar towing of implements and for the belt driving of machines such as threshers and balers.

While the Advance-Rumely Co. continued to refine its own Oil Pull line of tractors, it acquired Aultman Taylor in 1924 but was itself acquired by Allis-Chalmers in 1931. In this year Advance-Rumely marketed the Model 6A tractor, which was a modern looking machine for its time. It was powered by a six-cylinder Waukesha engine and fitted with a six-speed gearbox, Allis-Chalmers marketed the 6A only until existing stocks had been used up and less than a thousand were made.

Caterpillar

During the late nineteenth century Benjamin Holt and Daniel Best experimented with various forms of steam tractors for use in farming. They did so independently, with separate companies, but both were pioneers with track-type tractors and gasoline-powered tractor engines. Paralleling the developments of the steam excavator were experiments with tracked machinery referred to as 'crawlers'. Crawler technology would later diverge into separate and distinct strands of activity although the technology employed was essentially the same. One of these strands of activity is, of course, the crawler's agricultural application. Holt bought out Daniel Best in 1908 but later had to compete with Best's son C. L. 'Leo' Best.

The initial experiments involved wheeled steam tractors which were converted to run with tracks. The first test of such a machine took place in November 1904 in Stockton, California where a Holt Steam tractor had been converted to run on tracks. This had been accomplished by the removal of the wheels and the rear's replacement with tracks made from a series of 3in x 4in wooden blocks bolted to a linked steel chain which ran around smaller wheels, a driven sprocket and idler

on each side. Originally the machine was steered by a single tiller wheel, although this system was later dropped in favour of the idea of disengaging drive to one track by means of a clutch which slewed the machine around. From there it was but a short step to gasoline-powered crawlers, one of which was constructed by Holt in 1906. By 1908 28 Holt gasoline-powered crawlers were engaged in work on the Los Angeles Aqueduct project in the Tehachapi Mountains, something Holt saw as a proving ground for his machines. By 1915 Holt 'Caterpillar' track-type tractors were being used by the Allies in World War I.

During the 1910s and 1920s there was a considerable amount of litigation involving patents and types of tracklayers—Best and Holt were the two companies most frequently named in the litigation. Holt's patent for tracklayers left him in a position to charge a licence fee to other manufacturers of the time including Monarch, Bates and Cletrac. World War I intervened and much of Holt's production went to the US Army while Best supplied farmers.

After the war, the two companies competed in all markets and neither had a significant advantage over the other. In 1921 the Best Co. introduced a new machine, the Best 30 Tracklayer. This crawler was fitted with a light duty bulldozer blade was powered by an internal combustion engine and had an enclosed cab. Eventually, in 1925, Holt and Best merged to form the Caterpillar Tractor Co. Later that year the new company published prices for its product line; The Model 60 sold for $6,050, the Model 30 for $3,665 and the two-ton for $1,975. The consolidation of the two brands into one company proved its value in the next few years, the prices of the big tracklayers were cut, the business increased and sales more than doubled.

Cletrac

The history of another crawler maker is not so straightforward. Rollin H. White was one of a trio of brothers who established the White name in the US auto industry. In 1911 he designed a self-propelled disc cultivator which, although it never went beyond the experimental stage, established his interest in agricultural machines. He then worked on ideas for a crawler tractor with a differential system that allowed the machine to be steered with a steering wheel rather than the more usual system of levers.

His company was based in Cleveland, Ohio and known as the Cleveland Motor Plow Co. The name was changed in 1917 to the Cleveland Tractor Co and then changed again in 1918 to Cletrac. Cletrac's Model R was the first tractor to offer the newly-developed controlled differential commercially. The differential slowed drive to one track and increased it to the other. It was effective and became standard on Cletrac tractors, and later found wide acceptance in crawler technology. Other tractors from Cletrac were the models H and W. The Model F followed in 1920, and was made until 1922 as a low cost crawler tractor available in high clearance row-crop format. It was powered by a four-cylinder side-valve engine that produced 16hp at 1,600rpm at the drawbar. Its tracks were driven by sprockets mounted high on the sides of the machine which gave the tracks a distinctive triangular appearance The Model F retailed for $845 in 1920.

Previous page: Agricultural steam engines were physically large and started a trend for large sized tractors that endured even when alternative sources of power came along.

Above left: The International Harvester Titan was unveiled in 1919. It produced 22hp. IH also made the mighty Mogul, a 45hp tractor.

Left: Advance Rumely manufactured the Oil Pull line of tractors.

Case

The J. I. Case Threshing Co. was formed in 1863 to build steam tractors. Its first tractor appeared in 1892 and from here the company went on to become one of the leaders of the industry. As with so many companies Case built experimental gasoline-powered machines around the turn of the first decade of the 20th century—Case built its vehicle in 1911 and in the same year the massive Case 30-60 won a first place in the Winnipeg, Canada tractor trials. It weighed almost 13 tons (11,700kg) but found a market as it was made until 1916. A smaller version, the 12-25 was made from 1913 onwards, but it was the company's 20-40 model that won significant recognition at the Winnipeg trials in 1913.

Case experimented with smaller machines and produced the four-wheeled 9-18 model in 1916 to compete with the popular Fordson. It was the 9-18 tractor that in many ways established Case as a major manufacturer and more than 6,000 of the two versions, 9-18A and 9-18B were made by 1919. The 9-18 was a lightweight tractor designed to weigh around the same as a team of horses and capable of pulling a plough or driving a thresher.

All these early Case tractors were powered by flat twin engines-horizontally opposed twin cylinder engines-of varying displacements and used other of their components from the then current Case range of steam engines. It was generally accepted that the future was in smaller tractors that had more in common with automobiles than steam engines. Case experimented with small-sized machines and developed the three-wheeled 10-20 Crossmotor tractor. This was powered by a vertical in-line four-cylinder engine mounted transversely across the frame. It had a single driven wheel and an idler wheel on the rear axle. The driven wheel was aligned with the front steering wheel and the machine was capable of pulling a two tine plough. Between 1915 and 1922 approximately 5,000 10-20s were produced.

In 1917 the 10-18 was launched, it was similar to the 9-18 but featured a cast radiator tank and an engine capable of higherrpm. During the 10-18s three-year production run around 9,000 examples were made.

The 15-27 was a tractor designed for a three-tine plough and the first Case tractor to have a PTO (power take off) fitted. Its capabilities matched the requirements of the market to the extent that more than 17,500 were sold between the 1919 introduction and 1924 when it was superseded by the Model 18-32.

Case had not abandoned production of larger tractors altogether and offered the 22-40 between 1919 and 1924 and the 40-72 between 1920 and 1923. A mere 42 examples of the latter tractor were made. Each weighed eleven tons and when tested in Nebraska in 1923 produced a record 91 belt horsepower but used fuel in huge quantities. Case was an extremely strong company at this stage—as early as 1923 it had made 100,000 tractors: it would get stronger still.

A new president, Leon R. Clausen (1878–1965), was appointed to head the company in 1924. Clausen had been born in Fox Lake, Wisconsin and in 1897 had graduated from the University of Wisconsin with a degree in electrical engineering. He had experience of the tractor industry as a former employee of John Deere, an antipathy towards trade unionism and some disdain of customer demands believing that product design should be solely the province of the engineering department. He started the company working on a redesigned range of tractors. Under Clausen progress and innovation were cautious and conservative which in some instances allowed competitors to benefit

at Case's expense. One area in which Clausen wanted to advance the company was to offer a full line of implements to complement the tractor range. To this end Case bought out Emerson-Brantingham in 1928

Emerson-Brantingham was one of the pioneers of American agricultural machinery manufacturing with roots that stretched back to John H. Manny's reaper of 1852. The company, which was based in Rockford, Illinois, had purchased the Gas Traction Co. of Minneapolis in 1912 and become heavily involved in the manufacture of gasoline-engined tractors. In its range were the Big 4 Model 30, a 30-drawbar horsepower 10-ton machine which was subsequently enlarged into the Big 4 Model 45 which was rated at 45 belt horsepower and 90 drawbar horsepower but was even heavier. The model 20 of 1913 was a much smaller machine, and it was followed by the Model AA 12-20 of 1918, which was refined, in turn, into the Model K of 1925.

Emerson-Brantingham was huge company but was facing financial difficulties in 1928 when Case bought it. This acquisition gave Case a boost as it acquired valuable sales territory in the heart of America's corn belt through an established dealer network as well as an established line of farm implements. Case later dropped the Emerson-Brantingham line of tractors but retained many of their implements.

Other smaller companies experimented and innovated but were never realistic long term propositions, these included Ebert-Duryea, Lang, Fagiol, Kardell, Michigan, Utility and Happy Farmer.

In 1929 the Case Model L went into production based around a unit frame construction. It was notable because for the first time in 15 years the engine was not mounted transversely but longitudinally. The angular Case Model L was a great suc-

Above: A preserved early example of the Fordson tractor photographed in Florida

cess and was to remain in production until 1939 when it was replaced by the Model LA, a restyled and updated version of the L.

John Deere

Today, John Deere is the last of the early tractor manufacturers to have its founder's full name as its brand name. The John Deere colours of green with yellow details remain, too, as vibrant as ever. The history of the massive John Deere company starts in 1837. In that year John Deere, a 33-year-old pioneer blacksmith from Vermont, designed and made a 'self-polishing' steel plough in his small blacksmith's shop in Grand Detour, Illinois. He made it from the steel of a broken saw blade and found that the plough was capable of slicing through the thick, sticky prairie soil efficiently without becoming stuck or forcing itself out of the ground. As it cut the furrow it became polished and ensured that the soil would not stick. It was a major breakthrough in farming technology and became the first commercially successful steel plough in America. The plough was fundamental in opening the American Midwest to agriculture and ensuring high levels of crop production. With a succession

Above: Many early tractors, especially those from Ford, used proven auto-mobile parts adapted to agricultural use.

Right: Rumely tractors were manufactured in LaPorte, Indiana, USA.

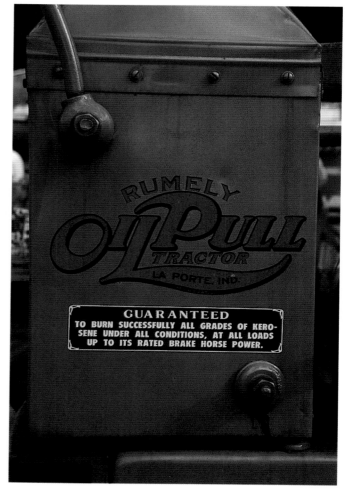

of partners John Deere made an increasing number of ploughs each year.

The supply of broken saw blades was, of course, limited so Deere had steel shipped from England. Then, in 1846, the first plough steel ever rolled in the United States was made to order for John Deere in Pittsburgh. In 1848 John Deere moved his operations to Moline, Illinois, to use of the waters of the Mississippi River for power to run his factory's machinery and for distribution of his products. By 1852 Deere and partners Tate and Gould were making approximately 4,000 ploughs per year. This partnership did not last long, in fact Tate later became a competitor, and by 1856 John Deere's son Charles was working at the company. In 1868 the company was incorporated as Deere and Co. Charles Deere went on to expand the company established by his father. He was reputed to have been an astute businessman who established marketing centres, called branch houses, to serve the network of independent retail dealers. During the 1880s John Deere was able to offer financing to retail customers for its agricultural products. By the time of Charles Deere's death in 1907, the company was making a wide range of steel ploughs, cultivators, corn and cotton planters, and other implements.

In 1911, under Deere and Co.'s third president, William Butterworth, six farm equipment companies were acquired and incorporated into the Deere organisation, going a long way to establishing the company as a manufacturer of a complete range of farm equipment. In 1918, the company purchased the Waterloo Gasoline Traction Engine Co. in Waterloo, Iowa, and this put John Deere in the tractor business. Tractors, of course, went on to become one of the most important parts of the company business and meant that the modern era of John Deere had begun.

The Waterloo Co. had produced the forerunner of what became its Model R by 1914. It was powered by a horizontal two-cylinder engine and was soon superseded by the Model N. This went on sale in 1916 and over its eight-year production run around 20,000 were made. The Waterloo Boy Model N was sold in Britain as the Overtime by the Overtime Farm Tractor Co., which was based in London. The Model N had a massive chassis frame on which was mounted the fuel tank, radiator, twin-cylinder engine and driver's seat. The engine, which was started with petrol and then run on paraffin, produced 25hp. The use of roller bearings throughout the machine was considered innovative at the time of its manufacture.

This early tractor had differing but equally important influences on two other tractor companies. The Belfast, Northern Ireland agent for Overtime tractors was Harry Ferguson. This was his first experience with tractors and started him thinking of better ways of attaching implements. It was also the Waterloo Boy Model N that brought John Deere into the tractor business.

In 1918 Deere and Co. bought the Waterloo Gasoline Engine Co. for a million dollars. At the time of this acquisition, the Waterloo Gasoline Engine Co. was making their Model N tractor and under John Deere supervision this model was kept in production until 1924. The Model N was rated as a 12-25 model when the tests confirmed outputs of 12.1 drawbar and 25.51 belt horsepower.

During the 1920s John Deere introduced its two-cylinder Model D tractor. Many other tractor makers were offering four-cylinder machines at the time but Deere's engineers, aware that the postwar boom in tractor sales was over, considered the economics of both manufacture and maintenance and designed the two-cylinder engine.

History showed that their decision was correct, as the Model D was a particularly successful machine. It was Deere's interpretation of the cast frame tractor powered by a two,cylinder kerosene-paraffin engine and fitted with a two forward and one reverse speed gearbox. From this basic machine a production run of a sequentially upgraded, tractors continued until 1953, by which time more than 160,000 had been made. The scale of upgrades can be gauged from he fact that the 1924 model achieved a rating of 22–30hp when tested in Nebraska and the 1953 model later achieved 38–42hp in the same tests. The distinctive exhaust note made by these two cylinder machines earned them the nickname of 'Johnny Poppers'.

John Deere brought out a row-crop tractor—the GP—in 1928, the same year that Charles Deere Wiman, a great-grandson of John Deere, took over direction of the company. During the period, when modern agriculture was developing, his strong emphasis on enginering and product development resulted in rapid growth. The GP designation indicated 'General Purpose' and it was designed as a three-row, row-crop machine and was the first John Deere tractor with a power lift for raising attached implements. Sufficient crop clearance was achieved by having a curved front axle and step-down gearing to the rear wheels. Unfortunately it did not prove as successful as had been hoped, so John Deere's engineers went back to the drawing board and produced the GPWT. The WT part of the designation indicated 'Wide Tread' and the machine was of a tricycle configuration. In subsequent years a number of variants of this model were produced including models for orchard use and potato farming, the GPO and GP-P models.

Eagle

The Eagle Manufacturing Co. was based in Appleton, Wisconsin and started tractor manufacture in 1906 with a horizontally opposed twin-cylinder engine-powered machine. By 1911 the company had in production a four-cylinder 56hp machine, and by 1916 was making a range of four-cylinder tractors that stayed in production through the 1920s.

Right: Early Fordson tractors were made in three countries in the early years, America, Ireland and England. Irish and English ones were exported to the US.

Fordson

In 1907 the Ford Motor Co. built the prototype of what it hoped was to become the world's first mass-produced agricultural tractor. The machine was based around components from one of Ford's earliest cars—including the transmission. As Henry Ford had grown up on a farm owned by his father, he was aware of the labour-intensive nature of farm work and keen to develop mechanised ways of doing things. As a result Ford built another tractor in 1915 and by 1916 had a number of working prototypes being evaluated.

Ford's prototypes, seen in action at a tractor trial in Great Britain were considered sufficiently practical a proposition for the British government to request that they be put into production immediately to assist with winning the war. Ford was preparing to do exactly that when in the early summer of 1917 German bombers attacked London in daylight. The Government saw this new development in the war as a serious threat and wanted to turn all available industrial production over to the manufacture of fighter aeroplanes in order to combat the German bombers. Ford was asked if he could make his tractors in the United States instead. He agreed, and only four months later Fordson Model Fs were being produced.

While this was only a temporary setback, the delay and shift in production caused Ford another problem. His plans became public knowledge prior to the tractors themselves going into production and another company tried to pre-empt his success. In Minneapolis a rival Ford Tractor Co. was set up, using the surname of one of its engineers. This shortlived company was not a great success but it did deprive Henry Ford of the right to call

Left: Titan tractors were marketed by International Harvester and some were imported into Britain to assist in farm production during WWI.

4 7

his tractors Ford. He resorted to the next best thing, Henry Ford and Son, shortened to Fordson.

Ford's aim was to do for farmers, with an affordable tractor, what the Model T car had done for motoring in general. He aimed to be able to offer a two-plough tractor for as little as $200. With a number of staff Ford designed what would later become the Fordson Model F. One of its innovative features was that it was of a stressed cast iron frame construction. This frame contained all the moving parts in dust-proof oil-tight units which eliminated many of the weaknesses of early tractors. It had four wheels and was compact, which gave it an unusual appearance at the time when both three-wheeled and massive tractors were the norm.

The Fordson Model F, Ford's first mass-produced tractor, went into production in 1917. The British Army ordered 5,000 of these Fordson trac-

Left and Below: B. Garnett's 1918 International Titan 10-20 driving a thresher at the Masham, England traction engine rally in the summer of 1999.

tors for the war effort. Power came from an in-line four-cylinder gasoline engine that produced 20hp at 1,000rpm. A three-speed transmission was fitted with a multiplate clutch that ran in oil, and final drive was by means of a worm gear. The ignition utilised a flywheel-mounted dynamo to supply high tension current to the coil which was positioned on the engine block. The tractor retailed at $750—more expensive than Henry Ford had predicted, but the reputation the Model T car had earned ensured that the new tractors would sell in large numbers.

The Model F proved immediately popular and US sales increased exponentially from the 35,000 achieved in 1918. Ford produced tractors that were reliable and gradually incorporated refinements as technology advanced. As early as 1918 Fordson had a high tension magneto, a water pump and an electric starter. Unsurprisingly, by 1922 Fordsons were accounting for approximately 70% of all US tractor sales.

However, by this time the postwar boom in tractor sales had ended and sales overall had declined. Ford survived by cutting the price of his tractors but his major competitors, notably International Harvester, did the same and the competition became fierce. By 1928 International Harvester had regained the lead in sales and had achieved 47% of the market total.

This had an effect on Ford: production of tractors by Fordson at Dearborn ended in 1928 although production continued briefly in Cork, Ireland before being transferred to Dagenham in 1932. Production was restarted in the US during the 1940s and has been continued ever since.

Heer

The Heer Engine Co. of Portsmouth, Ohio produced a four-wheel drive tractor in 1912.

International Harvester

The history of International Harvester, and of the 'Cornbinder' nickname, goes right back to 1831. In that year Cyrus McCormick invented a reaper, which while officially referred to as the McCormick Reaper, soon became known as the 'Cornbinder'. The International Harvester Corporation was formed in 1902 through the merger of McCormick with Deering. By 1907 the company was producing auto-wagons, a very early light-duty truck. Things went well for the company and by 1912 buses and more substantial trucks were being made.

The first International Harvester tractors appeared in 1906, when the Type A gasoline tractor was marketed with a choice of 12, 15 and 20hp engines. The Type B soon followed and lasted until 1916. The company followed this in 1919 with the Titan, a 22hp machine, and soon was selling two ranges of tractors, the Mogul and Titan models—the former built in the Chicago, Illinois factory and sold by McCormick dealers and the latter built in the Milwaukee, Wisconsin factory and sold by Deering dealers. The Titan 10-20 was and was the smallest but most popular model in the Titan range. The Moguls were heavier and International Harvester was noted for production of the giant 45hp Mogul tractor during the second decade of the 20th century.

Production of the Titan 10-20 started in 1914 and lasted for a decade because the simplicity of the design ensured a reputation for reliability. The 10-20 was powered by a paraffin-fuelled, twin-cylinder engine of large displacement that achieved 20hp at only 575rpm. The engine was cooled by water contained in a cylindrical 40-gallon tank positioned over the front wheels.

The Titan 10-10 was assembled on a steel girder frame, was chain-driven and had a two-speed transmission. More than 78,000 were made and it was one of several models of tractor were to carry the company into the 1920s. The others produced in that decade included the 8-16, the 15-30 and the Farmall Regular. (See below. Farmall was one of the tradenames used by International Harvester during the 1920s.)

The International Harvester 8-16 was produced in the USA during World War I and proved popular. It had a distinctive appearance as a result of a sloping hood over the engine and radiator. Production commenced in 1917 and lasted until 1922. The design, although inspired by the International Harvester truck line, was old-fashioned at the time of its introduction as it was based around a separate frame and featured chain and sprocket final drive. However, the availability of a PTO (power take off) as an optional extra gave the 8-16 an advantage over its competitors as the tractor could be used to drive other machinery.

While International Harvester could not claim to have invented the PTO (it had been used on a British Scott tractor in 1904), it was the first company to have commercial success with the innovation. International Harvester made it a standard fitting on its McCormick-Deering 15-30 tractor introduced in 1921.

Things really took off for International Harvester after World War I when, in 1921, a line of new trucks, designated the 'S' series were put on sale. They were sold through farm equipment dealers and quickly caught on with farmers. They were painted in a distinctive shade of red and because of this and their small size became known as 'Red Babies'. The 'S' stood for Speed-truck as the Red Baby was capable of 30mph.

Things went well for International Harvester C throughout the 1920s; in 1929, for example, they

sold 50,000 trucks through 170 outlets. This level of production enabled the company to compete with companies such as Ford and Chevrolet. International's truck production was, of course, carried out alongside their tractor production.

As the initial post-war wave of prosperity passed in the early 1920s, Fordson cut its prices to keep sales up. Faced with this, International Harvester was in a position to offer a plough at no extra cost with each of its tractors sold. This had the effect of selling all International Harvester stock allowing them to introduce the 15-30 and 10-20 models in 1921 and 1923 respectively. These machines were to give Ford competition in a way that company had not experienced until then. The new International Harvester models were constructed in a similar way to the Fordson around a stressed cast frame but incorporated a few details that gave them the edge—a magneto ignition, a redesigned clutch and the built in PTO-power take off—and so set a new standard for tractors.

The International Harvester 10-20 of 1923 is reputed to have offered 'sturdy reliability' and power and went on to become an outstanding success in the US where it was generally known as the McCormick-Deering 10-20. It remained in production until 1942 and production totalled 216,000. The styling of the 10-20 owed a lot to the larger 15-30 and both models were powered by an in-line four-cylinder petrol and paraffin engine with overhead valves that was designed for long usage with replaceable cylinder liners. A crawler version of the 10-20—the TracTracTor—was unveiled in 1928 and was the first crawler produced by International Harvester. It was to become the T-20 in 1931.

In 1924 the company progressed further with the introduction of the Farmall, the first proper row-crop tractor. It could be used for ploughing and turn its capabilities to cultivation too. It was suitable for use along rows of cotton, corn and other crops. These were refined and redesigned for the 1930s. The Farmall Regular became one of a range of three models as the F-20, with the F-12 and F-30. These machines were similar but had different capacities. In 1929 the 15-30 had become the 22-36 and, subsequently, was replaced by the W-30 in 1934. The 10-20 had a long production run being made until 1939.

Massey-Harris
Formed in 1891 in Toronto, Canada from the merger of two competing companies who manufactured farm implements, Daniel Massey had been making implements since 1847 while A. Harris, Son & Co. was a competitor for the same market.

Minneapolis Moline
In the late 1800s and early 1900s the Minneapolis Steel and Machinery Co. was primarily a structural steel maker that produced thousands of tons per year. The company also manufactured the Corliss steam engine which served as a power unit for many flour mills in the Dakotas. In 1910 Minneapolis Steel and Machinery produced a tractor under the Twin City name. This was the Twin City 40. Its steam engine influences were obvious; an exposed engine in place of the boiler, a cylindrical radiator and a long roof canopy. The Twin City 40 was powered by an in-line four-cylinder engine and its larger cousin the 60-90 by an in-line six of massive displacement. The 90 suffix indicated 90hp at 500rpm. The machine also had seven feet diameter rearwheels and a single speed transmission which made it capable of 2mph. By the time of the outbreak of World War I, MSM was one of the larger tractor producers and diversified into the manufacture of tractors for other companies—one so manufactured was the Bull tractor.

For the Bull Tractor Co., MSM contracted to build 4,600 tractors using engines supplied by Bull. The acceptance that small size tractors were practical quickly changed the emphasis of the tractor manufacturing industry and threatened some of the old established companies. The Bull tractor was a physically small machine based around a triangular steel frame. It had only one driven wheel thereby eliminating the need for a differential, a single wheel at the front steered the machine and the third simply free wheeled. An opposed twin engine produced up to 12hp and the transmission was as basic as the remainder of the machine. It had a single forward and single reverse gear.

Initially the Bull tractor sold well but as the limitations and inherent faults within the machine's design and construction became apparent sales quickly declined and little was heard of the company after 1915. The Fordson was to be an altogether more practical version of the same idea.

Through the 1920s MSM produced Twin City tractors and used slogans such as 'Team Of Steel' and 'Built To Do The Work'. After 1929, this line was still produced in the old MSM Lake Street Plant under the Minneapolis- Moline Twin City tractor banner.

In 1929 three companies, MSM, Minneapolis Threshing Machine Co. and the Moline Implement Co. all merged and Minneapolis-Moline Power Implement Co. was the result. Among these companies assets were Twin City Tractors and Minneapolis Tractors, and in the wake of the merger

Above right: A restored Fordson tractor in Florida; note steel lugs on rear wheels to aid traction.

Right: The American-made Bates Steel Mule was of an unusual design relying on a single crawler track for propulsion.

came the rationalisation of the products and factories. The Twin City tractor range was chosen as the one to spearhead the M-M push into the market with production continuing in the Minneapolis factory. Initially these were Twin City tractors with the M-M name added but as the range evolved the Twin City brand name was reduced in prominence and Minneapolis Moline became the brand name.

In 1915 the Moline Plow Co. had purchased the Universal Tractor Co. of Columbus, Ohio. The product line was moved to Moline, Illinois and a new building was built for the production of the Moline Universal Tractor, a two-wheel unit design for use with the farmers' horse-drawn implements as well as newly developed Moline tractor-drawn implements. The Universal tractor was commonly referred to as the first row-crop tractor. It was equipped with electric lights and a starter, which was very advanced for its time. The 27hp Universal was made between 1915 and 1923.

After World War I, some automobile manufacturers were looking to produce tractors and the Moline Plow Co, was courted by manufacturer John N. Willys. Willys purchased Moline from the owners, the Stephens family, and subsequently his automobile company began producing the Universal tractor. Willys had as his partners in the tractor trade, George N. Peek, a farm equipment executive, and General Hugh Johnson. Willys produced the Moline Universal tractor into the 1920s. But as the tractor boom subsided, Willys withdrew from the Moline Plow Company and sold out to his partners. General Johnson became President and R. W. Lea became Vice President of Moline

Right: This Bates Steel Mule was made in 1916; it produces 30hp. This example is the only one in Europe and was imported into Ireland in 1916. It is now owned by E. Bainbridge.

Top and Above: Early machines such as the Fendt Dieselross and Lanz
Bulldog were powered by simple semi-diesel engines.

Plow Company. When they retired their associates took over and operated as the Moline Implement Co., which it remained until the M-M organisation was founded in 1929.

The Minneapolis Threshing Machine Co. began producing steam traction engines just west of Minneapolis, Minnesota in 1889 where the town of Hopkins was founded and flourished solely because of The company. In 1893, a Victory threshing machine and steam engine built by the company won several medals at the World Exposition in Chicago. By 1911 the Minneapolis Threshing Machine Co. was building tractors under the Minneapolis name and went onto build many large tractors until the M-M merger, but these tractors were better suited to sod-breaking than for row-crop applications. After building a couple of row-type tractors the company marketed its Minneapolis 17-30 Type A and Type B. These were cross-motor row-crop tractors and remained in production even after the M-M merger. During the 1920s, until the M-M organisation, the company's products were advertised as being part of 'The Great Minneapolis Line'.

Oliver

Charles Hart and Charles Parr were engineering students together at the University of Wisconsin and graduated in 1896. They formed a company in Madison, Wisconsin to build stationary engines. These were unusual in that they used oil rather than water for cooling which was an asset in areas that suffered harsh winters simply because oil freezes at considerably lower temperatures than water thereby avoiding damage to engines caused by freezing coolant.

Hart and Parr moved to Charles City, Iowa in 1901 and made their first tractor in 1902. The Model 18-30 followed it in 1903 with a distinctive rectangular cooling tower which then became a distinguishing feature of all Hart-Parr tractors for the next 15 years. The cooling system circulated the oil around vertical tubes within the tower and air flow through it was maintained by directing the exhaust gases into the tower. The engine was rated at 30hp and was of a two-cylinder horizontal design with a large displacement that operated at around 300rpm. The 17-30 was a further developed tractor from the company and was followed by numerous others. These included the 12-27 Oil King. Hart-Parr redesigned its machinery for the post-war boom. The 12-25 was one of the new models and featured a horizontal twin cylinder engine. Slightly later these tractors were offered with an engine driven PTO but the full import of this was not realised until later.

In 1929, Hart-Parr, Nichols and Shepard and the American Seeding Machine Co. all merged with the Oliver Chilled Plow Co. to form the Oliver Farm Equipment Sales Co. and began to design a completely new line of tractors. Oliver itself dated back to 1855 and had the name of its Scottish-born founder, James Oliver, who had developed a chilled steel plough. He had patented a process that gave the steel used in his ploughs a hard surface and tough consistency. Once the Oliver Farm Equipment Sales Co. was formed the original Oliver Co.'s prototype row-crop models Models A and B went into production. Oliver was the first company to use laterally adjustable rear wheels to suit the differing row spacings of different crops, but the other manufacturers soon followed. These tractors were powered by a 18-27 four-cylinder engine that was also fitted to the company's line of conventional tractors, the Oliver Hart-Parr standard models. These were available in Standard, Western, Ricefield and Orchard versions and built until 1937. With a choice of four- or six-cylinder engines these tractors later became the Oliver 90 during the late 1930s.

Europe

The history of tractor building in Europe took a different path to that in the United States. The economic conditions were different, in that labour was more plentiful and cheap in Europe than in the US so that the pressure for mechanisation was not as great. However, the outbreak of World War I in Europe was to have far reaching effects both in terms of the economics of farming and production of tractors. The decimation of the male population and the devastation of great swathes of countryside meant that in the post-war years it became apparent that the tractor and mechanisation of farming was here to stay, like it or not.

Austria

Steyr entered the tractor market in 1928 when it announced an 80hp machine of which only a few were made.

Belgium

In 1910 Werkhuizen Leon Claeys, which had been founded in 1906, built its factory in Zedelgem, to manufacture harvesting machinery.

Right and Below right: In the early days of tractor manufacture there was no established configuration of engine and drivetrain until Ford established the Fordson concept.

Czechoslovakia

The noted motor manufacturers Praga and Skoda built tractors from quite early on; their constituent companies had previously offered motor ploughs. Praga offered the AT25, KT32 and U50 models while Skoda offered a four-cylinder powered three-speed tractor designated the 30HT. It could be run on either Kerosene or a mixture of alcohol and gasoline known as 'Dynalkol'.

In the mid-1920s, in a post-war Czechoslovakia now independent from the Austro-Hungarian Empire, the two-cylinder Wikov 22 was made by Wichterie and Kovarik of Prostejov.

France

De Souza and Gougis were just two of the makes that were presented at a pre-World War I tractor trial at the National Agricultural College at Grignon, near Paris. At this event tractors undertook a variety of voluntary and compulsory tests. In the early days of European tractor production, however, France generally trailed the market and tractor production was basically only carried out by Renault and Austins assembled in France. The French imported American tractors and renamed them with more French sounding titles. Globes were, in fact, renamed Allis-Chalmers 10-18s, Czars were Bean Tracpull 6-10s and Le Gaulois was a Galloway Farmobile 12-20.

The priorities changed considerably after World War I devastated France. The French government

Left: Fordson tractors were made by the Ford Motor Company in Dagenham, England.

made a major effort to regenerate the rural economy and helped this by giving interest free credit to businesses in the farming sector. Mechanisation was regarded as completely necessary and tractor trials were instituted at Rocquencourt in the spring of 1920. These trials evaluated both domestic French and imported models. In the autumn of the same year further trials were held at Chartres and 116 tractors were entered coming from 46 manufacturers from around the world.

French auto maker Renault had made light tanks during the war and so had proven experience in crawler technology. They turned this towards agricultural machinery and came up with a machine powered by a four-cylinder petrol engine and utilising tank and commercial vehicle parts. The tractor was subsequently designated the GP. It used a four-cylinder 30hp gasoline engine and a three-speed transmission with a single reverse gear; steering was by means of a tiller. An upgraded version, the HI, went on sale in 1920. Renault then developed the HO, a wheeled version of the HI. These models were powered by a four-cylinder engine that produced 20hp at 1,600rpm and featured epicyclic reduction gears in the rear wheels. All these models of tractor were based around a steel girder frame and the engine was enclosed within a stylish curved bonnet that bore a close resemblance to Renault's trucks and cars of the time.

The PE tractor was introduced in 1922 and was considerably redesigned from the earlier models. Renault introduced the VY tractor in 1933 powered by a 30hp in-line four-cylinder diesel engine. It had a front positioned radiator and the engine was enclosed. This model was painted yellow and grey and became the first diesel tractor to be produced in significant numbers in France.

Peugeot offered the T3 crawler and Citroen made a crawler tractor suited for vineyard use.

This latter machine was superseded by the Citroen-Kegresse. The success of tracks offroad led to the development of half-tracks such as the vehicles from Citroen-Kegresse which were proven with expeditions to Africa. French expeditions made the first Sahara desert crossing between December 1922 and January 1923. Later another French team drove from Algeria in North Africa to the Cape of Good Hope in the south between November 1924 and July 1925. A third expedition in 1931 took French crews from Beirut to French Indochina (Vietnam).

Other French tractor manufacturers included Somua, Amiot, Dubois, Latil, Gerde d'Or, Delahaye, Mistral, RIP as well as licence-built Saundersons and French-built Austins. Despite this, the economic situation meant that as French currency devalued, horses remained a more viable means of agricultural production than tractors so with the exception of Latil's timber tractors, Citroen-Kegresse half-tracks and the Renault machines interest in tractor production largely declined.

Germany

In Germany, in the late 19th century, Adolf Altona built a tractor powered by a single-cylinder engine that used chain drive to the wheels. This machine was not wholly successful but progress would be made in this country as a result of Dr Rudolph Diesel's experiments with engines. Diesel's engines used high compression to ignite the fuel and was the beginning of what is now known as the diesel engine.

The company that became known as Deutz was among the pioneers of the internal combustion engine. Nikolaus August Otto was an early proponent of the internal combustion engine and in conjunction with Eugen Langen manufactured a four-stroke engine which the duo exhibited in Paris,

France at the World Exhibition of 1867. The pair formed a company, Gasmotoren Fabrik Deutz AG and employed the likes of Gottlieb Daimler and Wilhelm Maybach. The company introduced its first tractor and motor plough, considered to be of an advanced design, in 1907.

The German company Hanomag from Hanover was offering a massive six-furrow motor plough at the outbreak of World War I. It was, like other German machines of the time, intended for the plains of Germany. In the years after this war Hanomag offered a larger version with an eight-furrow plough and an 80hp four-cylinder gasoline engine. Later in the conflict the Germans were blockaded by the Allied navies by 1917

and forced to rely on increasing their own production of food. They seem to have been more reliant on horse ploughs than other European nations, but the production of motor ploughs by Lanz, Stoewer and Hansa-Lloyd continued. Lanz manufactured the Landbaumotor, Stoewer the 3S17 and 6S17 models and Hansa-Lloyd the HL18.

After the war, the German economy was in ruins, but nevertheless the tractor market was tempting enough to make Benz start building tractors in 1919 when it offered the 40 and 80hp gaso-

Below: Steel-lugged driven wheels were the norm until the advent of the agricultural pneumatic tyre.

ALLIS CHALMERS

ALLIS-CHALMERS
1918   "6-12"
S/N 10419
Owned & Restored By
Jack Gustafson
BYRON, ILLINOIS

line engined 'Land Traktors'. By the beginning of the 1920s there were numerous small tractor manufacturers in business, although the likes of MAN and Hanomag were more established and produced machines in larger numbers. In 1924 Ford raised the stakes when the Fordson Model F went on sale in Germany, meaning that German manufacturers had to compete.

There was one major difference between the American and German machines—fuel. The German companies such as Stock and Hanomag compared the Fordson's fuel consumption unfavourably against their own machines, which were moving towards diesel fuel. Lanz, for example, had introduced the Feldank tractor that was capable of running on poor fuel through use of a semi-diesel engine. Lanz later introduced the Bulldog, for which the company became noted. The initial Bulldogs were crude, the HL model, for example, had no reverse gear and the engine was stalled and run backwards to enable the machine to be used in reverse. Power came from a single horizontal-cylinder two-stroke semi-diesel engine that produced 12hp. From this the HL was gradually improved and became the HR2 in 1926. Lanz later progressed its Bulldog models including the Model T crawler, L, N and P wheeled models. These produced 15, 23 and 45bhp respectively. Imports were made into the UK and the machines were popular because of their ability to run on poor grade fuel including used engine and gearbox oil thinned with paraffin. Lanz was later acquired by John Deere.

The Benz-Sendling S7 of 1923 was the first diesel-engined tractor manufactured by Benz. It featured power from a 30hp two-cylinder, vertical engine. The machine itself was a three-wheeled tractor with a single driven rear wheel, although outriggers were supplied to ensure stability during use. A four-wheeled machine, the BK quickly superseded the S7, and in 1926 Benz and Daimler merged and used the name Mercedes-Benz.

In 1926 Deutz unveiled the MTZ 222 diesel tractor and the diesel tractor technology race was underway. In Germany the diesel engine changed the face of tractor manufacture, and at the beginning of the 1930s Deutz produced its *Stahlschlepper*—Iron Tractor—which was produced in a number of models including the F1M 414, F2M 317 and F3M 315 models with single-, twin- and three-cylinder diesel engines respectively. The smaller engines were started by an electrical mechanism while the larger displacement one started with compressed air. When running, the largest displacement model produced 50hp. By this time Deutz was selling its engines to other tractor makers, including Fahr with whom they would later merge. Ritscher was another company that used Deutz diesels in the construction of what was the only tricycle-type tractor built in Germany

In 1921 and 1922 the Munktell company offered a two-stroke two-cylinder hot bulb engine that produced 15–22hp and, as a result, was designated the Model 22. Its twin-cylinder engine was actually derived from a marine engine and used compressed air to start after the hot bulbs had been heated with an integral blow lamp.

Great Britain

Hornsby of Lincoln was licence-building tractors in the 1890s. Later Petter's of Yeovil, Albone and Saunderson of Bedford constructed tractor-type machines. A tractor trial was held in Great Britain as early as 1910.

Left: Allis-Chalmers had roots in agricultural machinery that stretch back before the development of the gasoline tractor.

Petter's produced its Patent Agricultural Tractor in 1903. Dan Albone was a bicycle manufacturer with little experience of steam propulsion and was therefore able to approach the idea of a tractor from a different viewpoint. He took ideas from the infant automobile industry and built a tractor which he named after a river near his home, the Ivel. His machine had only three wheels but was practical and suited to a variety of farm tasks. It was a success and production commenced, some were exported and the company no doubt would have become a major force in the industry except for Albone's untimely death in 1906. The company seemed to stall without his involvement and ceased production in 1916.

Other manufacturers were coming on the scene by this time and included Ransome's of Ipswich; Marshall and Daimler built machines and looked to export them, a Marshall machine was exhibited in Winnipeg, Canada in 1908.

Until the outbreak of the war, however, Britain was in the fairly luxurious position of having much of its food shipped in from an empire that stretched around the world. This meant that farming in Britain was somewhat depressed—something that changed when the war suddenly exerted massive demands on British farms and farmers. The years of war saw the need for massive number of horses and men for the army, while the German Navy's U-boats aimed to starve Britain into surrender. The coalition government of the time instituted policies that encouraged domestic food production, and these policies ranged from ploughing and cultivating land that had been allowed to stand fallow, to increasing the mechanisation of farming in an attempt to be more productive with fewer people.

Some of the prewar tractor producers had gone over to other war related work. Ruston

Hornsby of Lincoln, for example, was involved with tank experiments. Saunderson tractors were still in production and Weeks-Dungey entered the market in 1915, but the obvious answer to the need for mechanisation was to import tractors from America. Soon after the Austin Motor Co. offered the Model 1 Culti-Tractor—actually a Peoria from Peoria, Illinois. The Big Bull was marketed in Britain as the Whiting-Bull, and a Parret model was renamed the Clydesdale in an obvious reference to horse-drawn ploughing. Another tractor to become famous was the Waterloo Boy, sold in Britain as the Overtime by the Overtime Farm Tractor Co. based in London. The International Harvester Corporation marketed the models from its range that were considered most suited to British conditions—the Titan 10-20 and the Mogul 8-16. During the war and in its aftermath, as Henry Ford's production of tractors on both sides of the Atlantic illustrates, there was considerable common ground between the tractor industries in Europe and the US

In the end, the British government policy paid off, and by 1918 the wheat harvest exceeded that of 1916 by approximately 50% and the production of other crops including barley, oats and potatoes had also increased significantly. The importance of the tractor to this, and its usefulness on the land was recognized. Numerous small tractor companies were formed post-war. One example is the Glasgow tractor, built between 1919 and 1924 in the Scottish city of the same name. It was a three-wheeled machine arranged with two wheels at the front and a single driven wheel at the rear to

Right: Massey-Harris was a Canadian agricultural machinery manufacturer who later merged with Ferguson.

eliminate the need for a differential. The front wheels had ratchets in the hubs to eliminate the need for one there, too. The company that produced the Glasgow tractor was the DL Co. which had taken on the lease of a former munitions factory after the Armistice.

As well as small tractor producers, larger manufacturers tried to move into the agricultural market. Austin of England manufactured a tractor powered by one of the company's car engines. It sold well despite competition from the Fordson and stayed in production until the early 1920s. In Peterborough another machine named after its place of manufacture originated, powered by a complex Harry Ricardo designed engine (see below). Ruston of Lincoln and Vickers from Newcastle upon Tyne manufactured tractors, and Clayton made a crawler tractor. As in America, all struggled to compete with the volume, price and quality of Fordson tractor production.

A good example of a British manufacturer at this time is shown by Roadless. On March 4, 1919, the company name Roadless Traction Ltd. was registered by Lt-Col Henry Johnson. Born in 1877 and having attended the King Edward VII School in Birmingham before studying engineering at the Durham College of Science, from here Johnson gained work experience in the heavy industries of South Wales prior to the outbreak of the Boer War. This started in 1899 and Johnson volunteered for the army, although he was not selected because of defective eyesight. He found his way, nevertheless, to South Africa by working his passage on a cattle boat. Once in Capetown he was able to get seconded to a steam road transport company of the Royal

Engineers as a result of his experience with steam engines. This unit was responsible for towing howitzers and field guns as well as ammunition behind mostly Fowler steam engines. It was here in South Africa that Johnson was able to study the use of mechanised transport in offroad situations. The war ended in 1902 and Johnson remained in South Africa and married Mary Henrietta in Bloemfontein. The couple came to England in 1906 when Johnson took up employment with the Leeds, West Yorkshire based Fowler and Company who required him to assist in their exports of steam engines. This involved a period of living in India where Johnson is reputed to have delivered many Fowler engines under their own steam to even the remotest parts of the sub-continent.

World War I started in 1914 and Johnson returned from India in 1915 to take up a post with the Ministry of Munitions. He spent much of those war years working on tank development and went out to France to the Front once the tanks were being used in combat. His tank development work, including that with rubber tracks and a spring and cable suspension system, continued after World War I. When Johnson's employment by the military ended he started converting Foden and team lorries to half-tracks by substituting the driven rear wheels with tracks and bogies. It is estimated that the company invested £50,000 in the development of its products in the seven years from 1921, and moved into Gunnersbury House, a former nunnery, in Hounslow, Middlesex in 1923. Johnson bought the house and leased part to the company and lived in the remainder. Motor lorries of varying sizes were also converted to half-track in this period including those from Peugeot, Vulcan, Austin, Guy, Daimler, FWD, and Morris Commercial. These were sold to customers in places as diverse as Scotland, Sudan and Peru. One

Left: Numerous tractors were made by the various companies that eventually became known as Minneapolis-Moline.

of the company's first major commercially successful orders was for a batch of lorries for the Anglo-Persian Oil Company for use in connection with oil exploration in Iran. Roadless supplied converted Morris Commercial lorries to fulfil this order.

Following on from this growing success the company turned its attention to tractors and potential agricultural applications for its technology. One of the first converted was the Peterborough tractor manufactured by the Peter Brotherhood Ltd. This tractor used a four-cylinder petrol-paraffin Ricardo design of four-cylinder engine that produced 30hp. Few of these tractors were made and even fewer—possibly only one—was converted to half-track configuration using the Roadless components. Roadless also co-operated with another Peterborough, Cambridgeshire based firm, Barford and Perkins Ltd, to produce a half-track tractor. This was based on that company's THD road roller and powered by a rear-mounted vertical two-cylinder McLaren-Benz engine driving through a three-speed forward and reverse transmission. It utilised Roadless tracks and had a drawbar pull rated at 15,120lb, but as it weighed 11 tons it was of limited use offroad and the project was abandoned.

The development of rubber-jointed crawler tracks by Roadless enabled the company to produce viable agricultural machines. The company's system—known as 'E tracks'—was easily adapted to many tractors and required little maintenance, so endearing it to farmers. The company converted AEC-manufactured Rushton tractors to full-tracks using E3 rubber-jointed tracks that were skid-steered by differential brakes. (Rushton Tractors was formed as an AEC subsidiary in

Right: A preserved Massey-Harris GP from 1932. It produced 22hp.

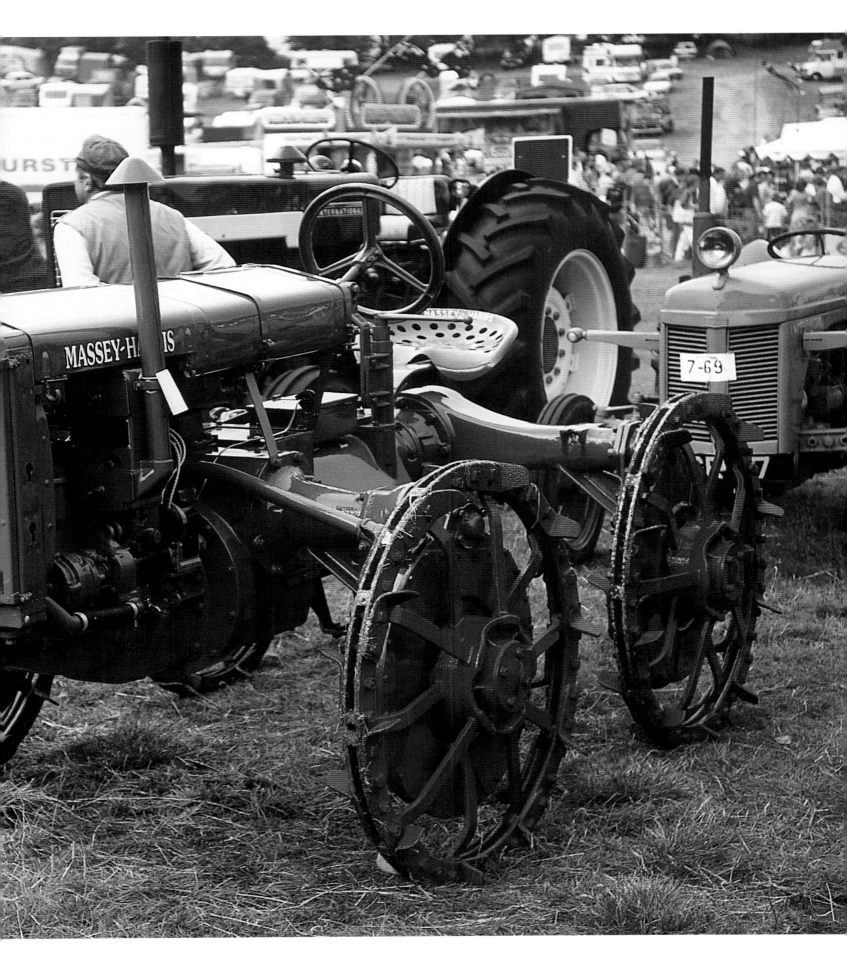

1929.) There were two variants of the Roadless Rushton with different lengths of track; the standard version had two rollers on each side while the other had three. These tractors were among those successfully demonstrated at the 1930 World Agricultural Tractor Trials held at Wallingford, Oxfordshire. The tractor was a success and sold in both the UK and export markets, although Rushton went out business after an Algerian customer defaulted on payment for 100 tractors shipped there for use in vineyards.

Hungary

Hofherr, Schrantz, Clayton and Shuttleworth (HSCS) was formed in Hungary after Clayton and Shuttleworth's withdrawal from the steam traction engine market in that country. Initially the compa-

ny's headquarters was based in Kispest, but it was later moved to Budapest. The company started making petrol engines in 1919, which led to the production of the company's first tractor in 1923. This first tractor used a single-cylinder petrol engine for its motive power and was assembled around a steel frame. Production versions of the HSCS tractor used single semi-diesel engines, a configuration that was popular in much of Europe but not in Britain or the US. The perceived advantages of the semi-diesel engine are that it is mechanically simple and will run on almost any type of fuel including waste oil. The HSCS engine was rated at 14hp and intended for ploughing and as a source of stationary power. The company persevered with semi-diesels throughout the production of numerous wheeled and crawler tractors.

Ireland

In 1919 production of the Fordson tractor had commenced in Cork, Ireland. This was the first model of tractor being manufactured simultaneously in the US and Europe. Production of Fordson tractors by Ford in the US ended in 1928 in the face of major competition from International Harvester but elsewhere, including Ireland and later England, it continued. In 1929 all Ford's tractor manufacturing was transferred to Ireland, making the Cork factory the only plant to be producing Ford tractors

Italy

Before World War I, there were a number of Italian tractor manufacturers, including Pavesi with its Tipo B. As with so many countries involved in World War I, Italian farmers began to realise the value of tractors in increasing productivity. The motor industry undertook experiments with tractors of its own design, and by 1918 the Fiat company had produced a successful tractor known as the 702 and Pavesi an innovative four-wheel drive and articulated steering tractor. In 1919 the first mass-produced Fiat tractor, the 702, came off the assembly line.

In 1911 Giovanni Landini had built a portable steam engine and from here he progressed to semi-diesel engined machines. After the war, Landini started work on his own design of tractor. His death in 1925 prevented the completion of the first prototype Landini tractor, but his sons took over the business and saw the completion of the tractor project, when, in 1925, the three sons built a 30hp machine.

Left: Caterpillar used weight designations for its early crawlers which were intended for agricultural applications.

Next page: The arched front axle seen on this John Deere was intended to increase crop clearance

Below: Minneapolis-Moline Model V.

In Italy during the1920s Breda was active in tractor production and unlike many of its competitors produced gasoline-engined tractors. These were powered by four-cylinder engines of 26 and 40hp.

Above: A Port Huron engine taking part in a vintage tractor-pulling class at an event in California.

Sweden

Munktells is credited with having built the first tractor in Sweden, the BM 30-40 of 1913. It was powered by a two-cylinder Bollinder engine. The company also experimented with tractors whose engines were wood-burning, simply because fuel was a major consideration in countries where it had to be imported. Oil products had to be imported to Sweden and were, therefore, expensive so that a machine that ran on as cheap a fuel as possible including waste oil offered clear advantages. This was the main reason behind Avance's first semi-diesel engined machine. The engineers at Avance had considered the starting procedure of diesels in detail and developed the semi-diesel which ignited the fuel by its injection onto a red hot bulb in the cylinder head. The Avance tractor

offered this new technology in a machine that was otherwise quite dated, featuring both a chassis and tank cooling of the engine. By the end of the decade Avance was offering two-cylinder hot-bulb semi-diesel engined tractors of two capacities; the 18-22 and 20-30hp versions. Both utilised compressed air starters with glow plugs and batteries as additional extra cost options.

Other Manufacturers

Other European manufacturers included Hurliman and Burer in Switzerland and Kommunar from the USSR.

On the other side of the world the McDonald tractor was unveiled in 1908 in Australia. This was followed by machines from Jelbart, Caldwell-Vale, Ronaldson and Tippet.

2 THE 1930s

Events within the world economy took a serious downturn with the Wall Street, New York Stock Exchange Crash of October 24, 1929. The crash would lead to the Great Depression which lasted into the 1930s and would see an estimated 13 million Americans out of work by 1932. The economics of production and competition meant that the 1930s started on a different note to the previous decade. The Depression slowed innovation rather than eliminating it, but by the end of the decade the world was sliding inexorably towards war.

North America

Through the Great Depression of the 1930s, and despite financial losses in the first three years of that

Left: Steel wheels with steel lugs, as fitted to this John Deere, were used on tractors to aid traction in fields prior to the advent of the pneumatic tractor tyre.

decade, **John Deere** elected to carry debtor farmers as long as was necessary. The result was a loyalty among its primary customer base that extends across three generations to the present day. Despite the setbacks of the Depression, the company achieved $100 million in gross sales for the first time in its history in 1937, the year of its centennial celebration.

The Wall Street Crash of 1929 led to the end of the optimism of the post-World War I years, and many of the numerous small manufacturers with a partially proven product went to the wall. What would remain after the Depression was a few larger companies with fully workable tractors, and new models that reflected both the increasing use of technology and, indeed, the progress of technology. Indicative of this shift is the history of the Eagle tractor. In 1930 the **Eagle** Co. moved into the manufacture of vertical in-line six-cylinder

powered tractors. One of these was the 6A Eagle, a tractor intended for three- to four-plough use and powered by a Waukesha six-cylinder engine which accorded it 22 belt horsepower and 37 drawbar horsepower. The Eagle was made in several forms—the Model 6B was a row-crop machine and the 6C a utility—and stayed in production until World War II when production ceased and never resumed.

Right: The John Deere GP tractor in tricycle row-crop form. The GP designation indicated General Purpose, as can be seen from the engine cowl.

Below right: The Case Model L of 1929. This tractor used an in-line four-cylinder engine and was the first Case tractor not to have a transversely mounted engine.

Below: The John Deere Model GP was introduced in 1928 as a three-row-crop tractor—later other variants based on the design were offered.

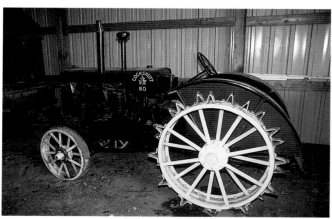

Above: The Cockshutt Plow Company of Brantford, Ontario, Canada, sold tractors in Canada including early Oliver models with the Cockshutt name. This Model 80 was one of their tractors from the 1930s.

Left: Steel lugged rear wheels and ribbed front ones as on the Cockshutt 80 were the norm before the advent of pneumatic agricultural tyres.

Below left: This tractor has large ribs bolted to its rear wheels, an alternative to the lugs. However, both types of attachment meant that tractors could not be driven on surfaced roads.

Right: The distinctive Case trademark was applied to the rear fenders of a tractor. The adjustable hitch that pre-dates the three-point linkage is seen clearly here.

Simple 1930s' developments such as the oil bath air filter gave engines a longer life when used in dusty conditions. The pneumatic agricultural tyre was the next step towards sophistication. The lack of pneumatic tyres had, until the early 1930s, hampered the widespread 'jack-of-all-trades' use of tractors, as those with lugged wheels could not be

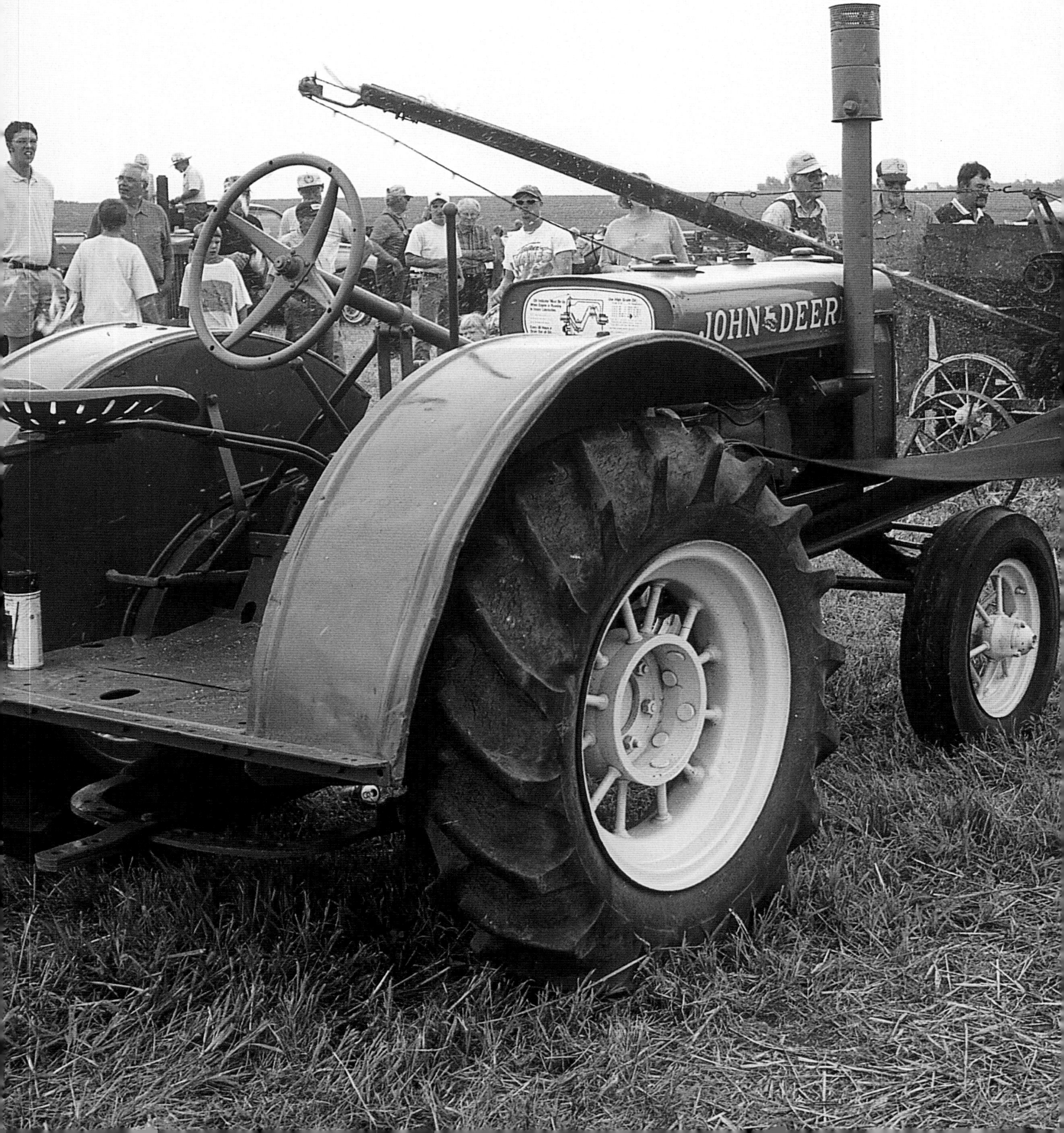
JOHN DEER

used on surfaced public roads, solid tyres suitable for road use were inadequate in wet fields and solid wheels and lugged wheels damaged crops and roots of crops. The tyre manufacturer B.F. Goodrich experimented with a zero pressure tyre, while Firestone experimented with modified aircraft tyres. These had angled lugs moulded on and were inflated to around 15lb/sq in (psi). The **Allis-Chalmers** Model U later became famous as the first tractor available with low pressure pneumatic rubber tyres.

In 1932 Model U tractors fitted with aircraft-type tyres inflated to 15 psi were successfully tested on a dairy farm in Waukesha, Wisconsin.

Left: The belt drive pulley located on the side of the tractor was important to farmers so that they could power machinery (as here), which is why tractors were given a horsepower rating at both belt and drawbar.

Below: A row-crop tricycle Case tractor with splined rear hubs that allow infinite adjustment to suit the spacing of rows of different crops.

Despite their proven ability pneumatic tyres were greeted with scepticism by those who thought that such tyres would not be adequate for farming use. Allis-Chalmers indulged in a series of speed events involving pneumatic-tyred tractors to promote this milestone in tractor technology. Allis-Chalmers was quick to embrace this breakthrough and went as far as hiring racing car drivers to display the company's new tractors with pneumatic tyres at speed at agricultural shows and State Fairs. One driver, Abe Jenkins, used an Allis-Chalmers Model U to break the world tractor speed record with a speed of 66mph. Alongside these developments were improvements in vehicle lighting and

Above: How the John Deere tractors looked before they were redesigned by Henry Dreyfuss in 1938 to give them a more up to date appearance.

Right: A 1931 John Deere GP model. The curved front axle was designed to give sufficient clearance over a growing crop.

Next page: The curved radiator grille seen on this Farmall row-crop tractor was fitted after Raymond Loewy redesigned the Farmall range during the 1930s.

fuel refining techniques that enabled improvements in efficiency and workability of tractors being achievable. The Allis-Chalmers WC with pneumatic tyres was introduced in 1934 as the first tractor designed for pneumatic tyres although steel rims were available as an option.

To complement these developments tractor manufacturers became aware of styling—at the time Ford was selling low priced but well-styled cars such as the V8 Model B. Allis-Chalmers introduced a bright orange paint colour to catch buyer's eyes, it was a simple ploy but one that no doubt worked. Other manufacturers soon followed suit with bright coloured paintwork and stylised bonnets, radiator grilles and mudguards.

The Model U and E tractors were the basis of the Allis-Chalmers range for much of the 1930s although in the latter years of the decade Allis-Chalmers introduced their Model A and B tractors. The four-speed Model A replaced the Model E and was made between 1936 and 1942 while the Model B ran between 1937 and 1957. The Model B was powered by a four-cylinder 15.7bhp engine and more than 127,000 were made. In 1936 the Model U was upgraded by the fitment of the company's own UM engine. The WC with pneumatic tyres was introduced in 1934 as the first tractor designed for pneumatic tyres although steel rims

were available as an option. In 1938 Allis offered the downsized Model B tractor on tyres and this was a successful move. The tractor was a sales winner and was widely marketed, it was manufactured in Britain after World War II for sale in both the UK and export markets. In this year the company increased the number of its tractor models with styled bonnets and radiator grille shells.

The **Massey-Harris** Model 25 was a popular tractor through the 1930s and 1940s. Another Massey-Harris tractor of this era was the 101 of 1935. This machine was driven by a 24hp Chrysler in-line six-cylinder engine. A third was the Twin Power Challenger powered by an I-head four-cylinder engine that produced approximately 36hp. The company chose a red and straw yellow colour scheme for its tractors in the mid-1930s. Nebraska Test No. 306 was on the Massey-Harris Model 101 S made by the Massey-Harris Co., Racine, Wisconsin. The dates between which it was tested were May 22 and May 26, 1939. The tractor's equipment consisted of 10-36 rear tyres, 5.00-15 front tyres, a six-cylinder L-head Chrysler T57-503

Left: The Allis-Chalmers Model U had a production run that spanned the entire decade. It was introduced in 1929 and production continued well into the 1940s.

Right: Dwayne Mathies at the wheel of his Farmall row-crop tractor on a South Dakota farm. It has been with the Mathies family since it was new.

engine run at 1,500 and 1,800rpm, 3.125in bore x 4.375in stroke, and an Auto-Lite electrical system. The tractor's weight was 3,805lb on steel wheels and 5,725lb on rubber tyres.

The Test H Data obtained with rubber tyres was as follows—

o   gear: 3
o   speed: 4.52mph
o   load: 1,987lb
o   rated load: 23.94 drawbar horsepower
o   fuel economy: 9.85hp hr/gal

The Test H Data on steel wheels was as follows—

o   gear: 2
o   speed: 3.68mph
o   load: 1,862lb
o   rated load: 18.29 drawbar horsepower
o   fuel economy: 7.46hp hr/gal

The fuel economy at 1,800rpm maximum with a load of 36.15 belt horsepower was 10.89hp hr/gal. Fuel economy at 1500rpm with a rated load of 31.5 belt horsepower yielded 11.86hp hr/gal

By 1937 the American economy was well on the way to recovery and in 1938 the **John Deere** company unveiled a range of tractors that had been redesigned by industrial designer Henry Dreyfuss. The reason for this was that the company's management was conscious that the John Deere line had changed little between 1923 and 1937 and many of their competitors were moving towards stylised tractors and taking their design cues from the auto makers of the time. The Models A and B were the first styled tractors, although mechanically the Model B was similar to the one originally introduced in 1935 and followed by the similarly styled D and H models in 1939 and the Model G in 1942. The Model H was a lightweight two-row tricycle tractor and over 60,000 were made.

Below: The Farmall F-14 replaced the Farmall F-12 in 1938. The F-14 was more powerful than its predecessor and rated at 14 drawbar horsepower in the Nebraska test.

Right: John Deere tractors were nicknamed 'Johnny Poppers' because of the distinctive sound made by their two-cylinder diesel engines.

Below right: A row-crop John Deere GP tractor hitched to a John Deere two-furrow plough.

The Model A was produced between 1934 and 1952. It was a tricycle row-crop tractor that incorporated numerous innovations. The wheel track was infinitely adjustable through the use of splined hubs, and the transmission was contained in a single-piece casting. The first Model A was rated at 18 drawbar and 24 belt horsepower but this output was subsequently sequentially increased. By the time production of the Model A was halted in 1952, in excess of 328,000 had been made.

The Model B was made between 1935 and 1952. It was smaller than the A and rated at 11 drawbar and 16 belt horsepower. Later it was produced in numerous forms, including the model BO of the 1940s and the crawler track equipped version of the late 1940s. The MC model made later was purpose built as a crawler tractor.

The Model G was a three-plough tractor rated at 20–31hp and the most powerful tractor in John

Above, Below and Right: The row-crop tricycle configuration became popular as indicated by these restored John Deere GP models. The later styled models with infinitely adjustable hubs at right are a Model B (**top**) and a Model H (**bottom**).

Deere's line-up at the time. By the time of its manufacture pneumatic tyres were supplied as the standard fitment although the rubber shortages caused by World War II would lead the maker to return to steel wheels.

The **Hart-Parr** company achieved unexpected success with the Oliver Hart-Parr Row Crop 70 HC that was introduced to the farming public in October 1935. This streamlined machine was fitted with a high compression, gasoline-fuelled six-cylinder engine and was noted as being quiet and smooth running. By February 1936 Oliver had sold in excess of 5,000 examples which was more than double what the company anticipated selling. In 1937 Oliver dropped the Hart-Parr suffix from its company name so that the Model 70 row-crop models carried only the Oliver name on their streamlined bodywork. In the Nebraska Test the gasoline-fuelled, six-cylinder row-crop 70 produced 22.64 drawbar horsepower and 28.37 belt

Above: Massey-Harris was a Canadian maker of tractors which later merged with Ferguson to form Massey-Harris-Ferguson. Subsequently the Harris portion of the name was dropped.

Above right: The Silver King tractors of the 1930s were noted as being fast machines and their model designations referred to the track of the rear wheels in inches.

horsepower. The Models 80 and 90 followed in 1938. The 80 had the angular look of earlier models while the 90 was run on kerosene. Oliver made both 90 and 99 models between 1938 and 1952. The former was a three-speed 49hp machine.

The Model C was a respected **Case** tractor of the era, it went into production in 1929 and, along with the Model L, was one of the tractors that helped establish the Case brand in Great Britain. It was one of the tractors tested in the 1930 World Tractor Trials in Oxfordshire where it achieved the best economy figures in the class for machines fuelled by paraffin. The Model C recorded a maxi-

mum output of 29.8hp on the belt and 21.9hp on the drawbar, figures that were almost identical to those achieved in Nebraska a year earlier. The results of the tests were widely advertised by **Case** who later offered the Models CC , CI, CO, CV and CD—row-crop, industrial, orchard, vineyard and crawler versions respectively.

Case bought the Rock Island Plow Co. in 1937. Following this acquisition, and that of the Showers Brothers furniture factory in Burlington, Iowa Case moved towards production of combine harvesters. A new line of tractors appeared in 1939, the Models D and DC which were to be followed by the Models S and V and an upgraded Model L, the LA. The Model D was a four-wheeled tractor and was designed to pull a three-tine plough, it had a belt pulley, PTO and Case's motor lift system for the implements. The DC was a row-crop tricycle with adjustable wheel spacing and what the manufacturers termed 'quick-dodge' steering for close

cultivation in uneven rows. The D-series were the first Case machines to be painted in a bright hue, Flambeau Red, a colour that was to become standard for the next decade.

The styling of the 1939 Case Model LA gave the tractor generally a more rounded appearance typical of the time but much of the engineering was similar to that of the earlier, more angular Model L. The engine was a 6.6-litre overhead valve in-line four-cylinder unit. Drive to the rear axle was by a pair of chains and sprockets as the Model L. The gearbox was a conventional four-speed unit with a single reverse gear and a lever-operated over-centre-type clutch allowed selection of the required gear. The tractor was successful and stayed in production until 1955.

International Harvester entered the 1930s optimistically, and a landmark was a redesign for 1934 that gave the trucks car-like styling and dropped its hitherto traditional grey paint scheme

PIONEER PARK DAYS
ANTIQUE ENGINE & CAR SHOW
ZOLFO SPRINGS, FLORIDA
MAKE 1936 SILVER KING
FATE-ROOT-HEATH-CO.
OWNER PLYMOUTH OHIO
BRIAN BENEDICT
COMPLIMENTS OF
HARDEE COUNTY
BOARD OF COMMISSIONERS

SILVER KING

around the same time and replaced it with red. The first wheeled, diesel-powered tractor appeared in 1934 designated the WD-40. It was powered by a four-cylinder engine and the numerical suffix to its power in the region of 44bhp. The engine was slightly unusual in that it was started on petrol and once warm switched to diesel through closing of a valve.

In the late 1930s the industrial designer, Raymond Loewy, redesigned International Harvester's trademark and then the appearance of the company's range of tractors. It was Loewy who gave this company's tractors the rounded radiator grille with horizontal slots. The new styling initially appeared on the crawler models but these were quickly followed by the restyled Farmall M tractors. This in turn was replaced by the K-series introduced in 1940.

Minneapolis Moline took the concept of styling further but after experiments with the luxury UDLX tractor with an enclosed cab launched

Above: The Farmall F-14 seen here was one of the last Farmall tractors produced before the new styled versions were introduced in 1939.

Left: Silver King tractors were made by Fate-Root-Heath of Plymouth, Ohio, a company who were originally makers of railroad locomotives but started tractor production in 1933.

Next page: A vintage Farmall tractor demonstrating corn harvesting near Hartford, South Dakota.

the GT as a five-plough tractor in 1939. In the Nebraska Tractor Tests its measured power output was 55hp at the belt, despite being rated at only 49hp by its manufacturer. Powered by a petrol-fuelled in-line four engine, it developed its maximum power at 1,075rpm. Later, in the aftermath of World War II, increased power versions the G and GB models were offered. The International Harvester W6 was announced in 1940 as part of a range of new tractors. The range also included the W4 and W9 models and all were powered by an in-line four-cylinder that produced 36.6hp when

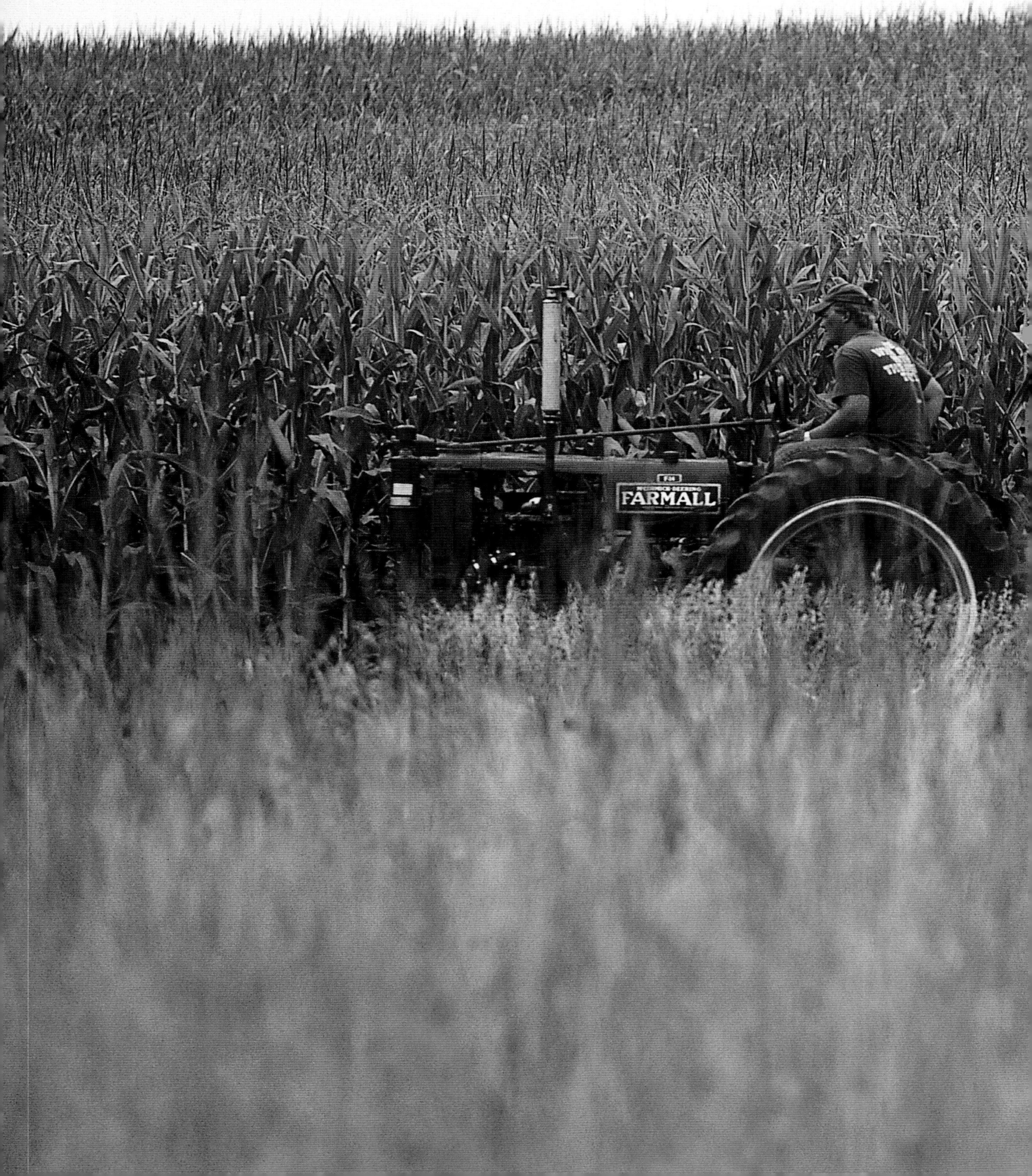
FARMALL

running on gasoline and slightly less on paraffin. A diesel version was designated the WD6.

Crawler tractor maker **Cletrac** introduced its first wheeled tractor in 1939, the general GG, a crawler version was also available and referred to as the HG. These models were powered by a Hercules IXA engine which was an in-line four that made 19hp at 1,700rpm at the belt. It was later upped to 22hp. The company also made a tricycle row-crop tractor, although production of this model was taken over by B.F Avery in 1941 from when it was sold as the Avery Model A.

Caterpillar stayed with full tracks; the Caterpillar 20 was a mid-sized crawler tractor put into production by the company in Peoria, Illinois late in 1927; production lasted until 1933. It was powered by an in-line four-cylinder engine that made 25hp at 1,250rpm. According to its makers it could pull 4,160lb at the drawbar in first gear, but

Above: For a period during the 1930s and early 1940s Henry Ford and Harry Ferguson co-operated in the production of tractors. These machines, as here, had both Ford and Ferguson badges above the grille.

Above right: Caterpillar adopted its now famous yellow colour scheme in 1931 and in this decade began to diversify from agricultural machinery production into construction machinery.

Right: The D2 Caterpillar crawler tractor went into production in 1938—it was the smallest Caterpillar crawler and rated at 19.4 drawbar horse-power in the Nebraska tests.

in the respected Nebraska Tractor Tests it recorded a maximum pull of 5,271lb. The transmission was a three-speed with a reverse and steering was achieved through multi-plate disc clutches and contracting band brakes. With a width of only five feet and length of nine, the tractor was compact and helped establish Caterpillar as a known brand on smaller farms and in export markets.

Above: A 1937 Fordson Model N. Tractors such as this were vital to the British war effort in that they facilitated increased agricultural production.

Above left: The International Harvester W-30 was introduced in 1934. It was later refined when the IH range was styled by Raymond Loewy.

Left: A 1931 25hp Fordson Model N; this would have been assembled in Ireland—although production moved subsequently to Dagenham, England.

In 1929 the company announced the Caterpillar 15 which increased the range of crawlers to five models of varying capabilities. In 1931 the first Diesel Sixty Five Tractor rolled off the new assembly line in East Peoria, Illinois, with a new efficient source of power for track-type tractors. Also starting in 1931 was the shift to the now familiar yellow colour for Caterpillar products; all Caterpillar machinery left the factory painted in a shade described as 'Highway Yellow'. New diesel-engined crawlers came in 1935 and model desig-

nations began RD—said by some to be Rudolf Diesel's initials—and were finished with a number that related to the crawler's size and engine power so there were RD8, RD7 and RD6 machines soon followed by the RD4 of 1936. Other accounts suggest that the R stands for Roosevelt, the D for Diesel and the 8 equates approximately to the machine's engine capacity. The RD8 was capable of 95 drawbar horsepower, while the RD7 achieved 70 drawbar horsepower and the RD6 45 drawbar horsepower.

By this time the US Forest Service was using machines such as the Cletrac Forty with an angled blade on the front, so Caterpillar built one with a LaPlante-Choate Trailblazer blade. Ralph Choate had started in business by building blades to be fastened to the front of other people's crawlers, his first one was used on road construction work between Cedar Rapids and Dubuque, Iowa.

Above: Fordson tractor production continued through the 1930s and almost unchanged into the war years; this is a 1942 25hp Fordson Model N.

In 1938 Caterpillar started production of its smallest crawler tractor, the D2. It was designed for agricultural use and capable of pulling three- or four-tine ploughs or a disc harrow. The Nebraska tests rated it as having 19.4 drawbar horsepower and 27.9 belt horsepower. A variation of the D2 was the gasoline or paraffin powered R2 which offered similar power output.

The next important innovation—one of the most important in the history of the tractor—was the introduction of the 'three-point hitch' system in the late 1930s. It was to be introduced on **Ford**'s 9N model having been designed by Harry Ferguson, an Irishman. Ferguson designed the first three-point hitch system that is used on farm tractors today. The ingenious system combined with compatible three-point implements, was a viable replacement for the horse and the horse-drawn implements. It also allowed the attachment of a variety of farm implements. Ferguson demonstrat-

ed the system to Ford in Michigan in the fall of 1938 and through the famous 'handshake agreement' by which each man's word was considered sufficient for the business partnership, production was to start.

At that time it was known as the Ferguson System and was produced in co-operation with the Ferguson-Sherman Company until 1946. The Model 9N was first demonstrated in Dearborn, Michigan, on June 29, 1939. This agricultural concept revolutionised farming. Many of the basic design principles and features incorporated into the 9N are still seen in tractors currently being manufactured. The Ford 9N, the first of the N Series tractors went on sale complete with the first three-point hitch in 1939. It was developed as a versatile all-purpose tractor for the small farm and was exceedingly popular. The 9N was powered by an in-line four-cylinder 120cu in (1966cc) displacement gasoline engine. Many of the engine's internal components including the pistons were compatible with parts used in Ford's automobile V8 of the time. The 9N went through subtle changes during every year of its three year production run. For example, in 1939 the grille had almost horizontal bars and the steering box, grille, battery box, hood, instrument panel and transmission cover were made of cast-aluminium. It also had clip-on radiator and fuel caps while in 1940 these caps were changed to a hinged type.

Europe

In **Great Britain** tractor trials were inaugurated in Benson, Oxfordshire, England in 1930. The first running of the event—the World Tractor Trials—

attracted a variety of interest from British, US and European tractor manufacturers, and from the 'international' Fordson at this time being made in Ireland. The British machines were produced by the likes of AEC Limited, Marshalls, Vickers, McClaren and Roadless. European manufacturers represented included Mercedes-Benz, HSCS, Latil and Munktell.

Below: Tractors competing in the 1930 World Tractor Trials in Oxfordshire, England: the French-made Latil (**top**) and the International Harvester Farmall tractor (**below**).

Ford's UK tractor production was established at Dagenham, Essex in 1932 and from there Fordsons were exported around the world including to the United States. Ford produced the All Around row-crop tractor there specifically for the US in 1936, for example. Production of Fordsons was not restarted in the US until the 1940s. The Fordson was the first foreign tractor tested in Nebraska at the noted Nebraska tractor tests. It was submitted for testing in both 1937 and 1938.

As the Fordson tractor gained popularity, Roadless Traction turned its attention to the up and coming machine. In 1929 the first Roadless conversion to a Fordson tractor was carried out on a Fordson Model N. It was successfully demonstrated on Margate Beach in early 1930 when the tractor proved capable of hauling three tons of seaweed off the beach. Roadless-converted Fordsons soon became popular and were offered

in two track lengths. Ford approved the conversion for use with its tractors and the association between Ford and Roadless Traction would endure from then until the 1980s.

Another major tractor maker that took advantage of the Roadless crawler technology was Case of Racine, Wisconsin. A variety of Case tractors—including the Models C, L and LH—were converted and, unlike the other conversions, the Case tractors retained the steering wheel. Roadless built experimental machines for McLaren, Lanz, Bolinder Munktell, Mavag and Allis-Chalmers, although many never progressed beyond prototypes.

Crucially important to the development of the tractor—as has already been discussed—was the introduction of the three-point system in the late 1930s. The David Brown-manufactured Ferguson tractors of the mid-1930s came about as a result of Harry Ferguson approaching them with regards to a tractor transmission. Brown's were noted for making gears and Ferguson wanted to produce a tractor with an American Hercules engine, and an innovative hydraulic lift. These machines were to

These pages: Tractors competing in the 1930 World Tractor Trials in Oxfordshire, England: the Aveling and Porter high speed diesel (**above left**), an Allis-Chalmers (**below left**) and (**below**) a diesel-powered Mercedes-Benz. Unsurprisingly ploughing was a major part of the trials.

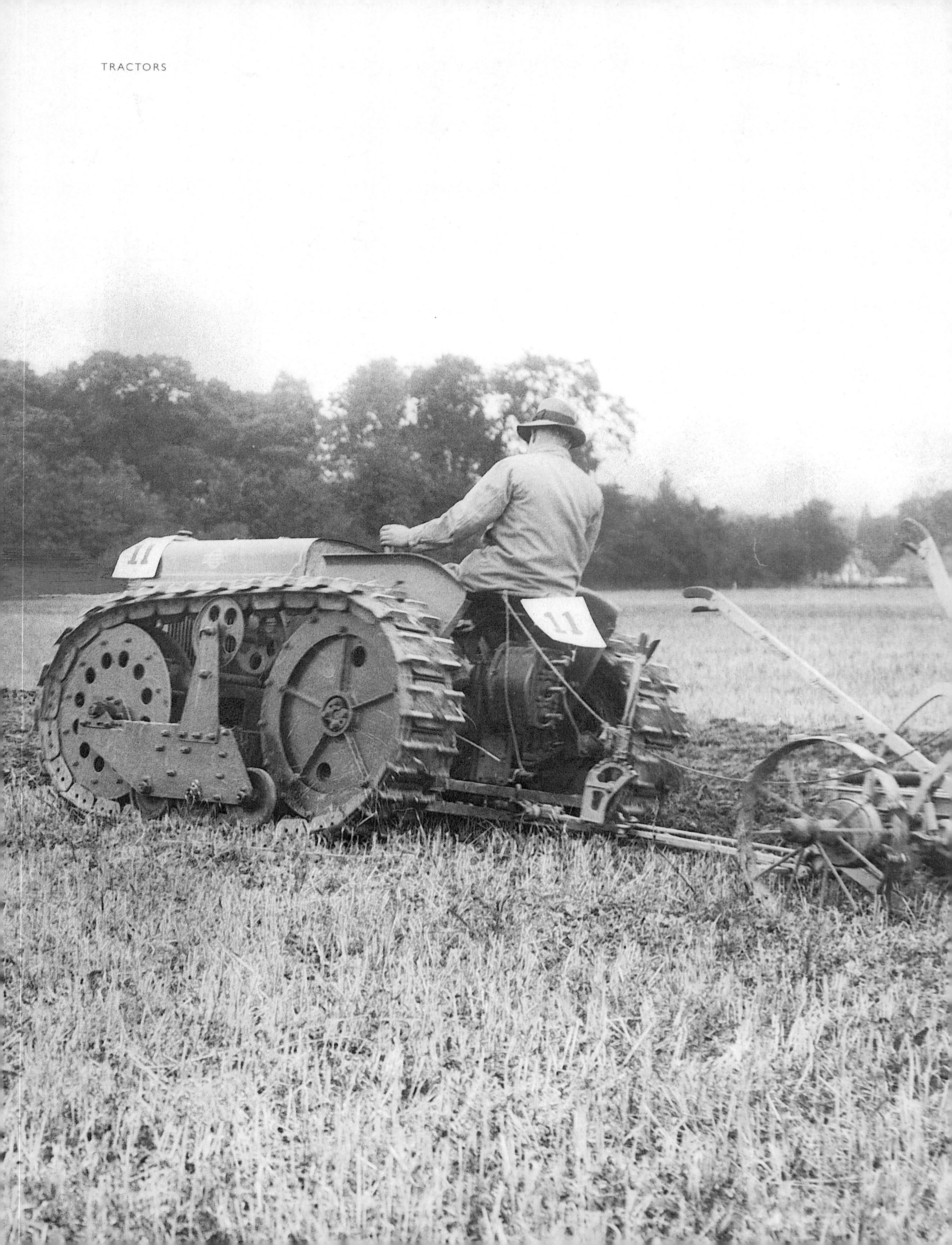

be built in Huddersfield, England by Brown after the prototype had been tested. It was called the Ferguson Model and fitted with a Coventry Climax engine and, latterly, an engine of Brown's own design. Production ceased in 1939 because of the divergence of the makers' aims; Brown wanted to increase power and Ferguson to reduce costs. While Harry Ferguson went to the US to see Henry Ford, David Brown exhibited a new model of tractor to his own design. The new machine was the VAK-1 and featured a hydraulic lift.

France generally still trailed the market somewhat; its tractor production basically the province of just Renault and Austins made in France. Renault introduced the noted VY in 1933 powered by a 30hp in-line four-cylinder diesel. It had a front-positioned radiator and enclosed engine. It was painted in yellow and grey and became the first diesel tractor to be produced in any significant numbers in France. Other products available included crawlers and specialist machines from the likes of Citroen-Kegresse and Latil as well as smaller vineyard tractors.

SFV—Société Français Vierzon—entered the market in 1935 with machine not unlike the Lanz Bulldog. Somua was another French manufacturer of agricultural machinery in the interwar years, and among its products was a machine with a rear, power-driven rotary cultivator.

Other French makers of similar machines included Amiot and Dubois. The former was a machine with an integral plough while the latter was a reversible motor plough.

In **Germany** the diesel engine changed the face of tractor manufacture. During the 1930s the Mercedes-Benz company produced the OE model which was powered by a horizontally arranged single-cylinder four-stroke diesel engine that produced 20hp. One of these Mercedes-Benz machines was entered at the 1930 World Tractor Trials held in England. Prior to World War II, tractor production in Germany was relatively small in number and even in the coming conflict the German Army was to rely heavily on horse-drawn vehicles in stark contrast to the mechanised armies of the Allies—at the end of the war the German Army still had more horses than vehicles.

By the 1930s Hanomag was thriving, partially as a result of exports of both its crawler and wheeled machines. In 1930 Hanomag offered the R38 and R50 wheeled models and K50 crawler; all were

These pages: Four-cylinder tractors competing in the 1930 World Tractor Trials in Oxfordshire, England: a Rushton Roadless crawler tractor with a petrol engine (**left**) and a kerosene (paraffin)-fueled International Harvester 10-20 (**right**).

diesel-powered and featured a PTO. Deutz produced the Stahlschlepper or Iron Tractor models, including the FIM 414 and F2M 317 with single- and three-cylinder diesel engines respectively. Lanz progressed its Bulldog models including the Model T crawler, L, N and P wheeled models; these produced 15, 23 and 45bhp respectively. Imports were made into the UK and the machines were popular because of their ability to run on poor grade fuel including used engine and gearbox oil thinned with paraffin.

One of the simplest tractors available in Germany during the 1930s was the Fendt Dieselross or 'diesel horse'. It was little more than a stationary engine equipped with a basic transmission system and wheels. During World War II, Fendt was among the German tractor manufacturers that offered gas generator-powered tractors that would burn almost any combustible material

as a result of the fuel shortages caused by the war. In the postwar years Fendt reintroduced its Dieselross tractors in different capacities. A 25hp version was powered by an MWM twin-cylinder engine while a 16hp model used a Deutz engine. Hanomag contributed to the German war effort and the company's factories suffered extensive damage. The company resumed production of tractors in the latter years of the 1940s and offered the four-cylinder diesel R25 tractor finished in red.

Above right: The Farmall F-14 (seen here) replaced the F-12 model in 1938. The F-14 was a 14–17hp tractor and stayed in production only until 1939.

Below right: The row-crop tricycle Farmall F-30 was made between 1931 and 1939.

Below: Case offered a new line of tractors in 1939 and followed this with an upgraded Model L—the LA seen here.

Fiat was still the major producer in **Italy** and by 1932 had begun production of the first tractor which can be considered as an earth-moving machine—the 700C, a tractor equipped with a front blade to shift earth and welded devices to load trucks. The 700C and was powered by a 30hp four-cylinder engine. Later, but before the outbreak of war, Fiat produced the 708C and Model 40.

Breda continued in tractor production and Alfa Romeo experimented with a machine while other makers, such as Landini, Deganello and Orsi, produced Lanz machines under licence. A range of Landini 40 and 50hp models were made during the mid-1930s and were to become known as Velite, Buffalo and Super. The first production tractors were built powered by a 40hp semi-diesel engine, which was a two-stroke single-cylinder unit. The Landini brothers also designed a more powerful

Right: Ford's Dagenham, England, plant soon became a major producer of tractors as this 1939 picture shows.

Below: Ford also used a tractor as a load to demonstrate the ability of its 30cwt trucks on steep Devonshire hills in southwest England.

tractor which they named the Super Landini, and in 1934 it was the most powerful tractor on the Italian market. It produced 50hp at 650rpm through an engine displacement of 74.4cu in (1,220cc). The tractor was driven through a three-speed transmission with a single reverse gear and it stayed in production until the outbreak of World War II. During the 1930s Breda offered a range of models that included both a conventional four-cylinder gasoline kerosene-engined tractor and an unusual two-stroke two-cylinder Junkers diesel-engined machine. After World War II the company offered multi-cylinder engined crawler tractors.

Hurlimann remained in business in **Switzerland** and produced the 2M20 in 1934 with a two-cylinder engine that produced 20hp.

In **Czechoslovakia** the two-cylinder Wikov 22 was made by Wichterie and Kovarik of Prostejov.

Mavag entered tractor production in **Hungary** during the 1930s.

Claas began the development of harvesters suited to European conditions.

In **Sweden**, by 1930 Munktell was producing its 20-30 model and this and the smaller 15-22 Model 22 took part in the Oxfordshire, England World Tractor Trials of 1930. Another Swedish manufacturer was Bofors, the arms manufacturer who entered the tractor market alongside Avance and Munktells with its 40–46hp two-cylinder tractor of 1932. This was the same year that Munktell combined with its engine maker, Bolinder, to form Bolinder-Munktell and continued the production of the 15–22 and 20–30 tractors. Munktell also experimented with wood burning tractors simply because fuel was a major consideration in countries where fuel had to be imported. The Bofors company offered a 31hp twin-cylinder engined tractor in 1939.

The Depression slowed innovation rather than eliminating it , and did not deter new manufactures

Far left: February 1939; English-manufactured Fordson tractors awaiting shipment to Vancouver and Calgary in Canada.

Above: A Fordson tractor being shown to be ideally suited to orchard work at a demonstration run by the Ministry of Agriculture and Fisheries in Ledbury, Herefordshire, England.

Above left: Sir Malcolm Campbell with his Fordson tractor fitted with Firestone Golf Course tyres used on the speed record maker's own golf course in Horley, England, in August 1935.

Left: A photograph taken in July 1935 of the Golf Course Fordson tractor. It was suited to working a seven or nine-gang mower as well as rolling greens.

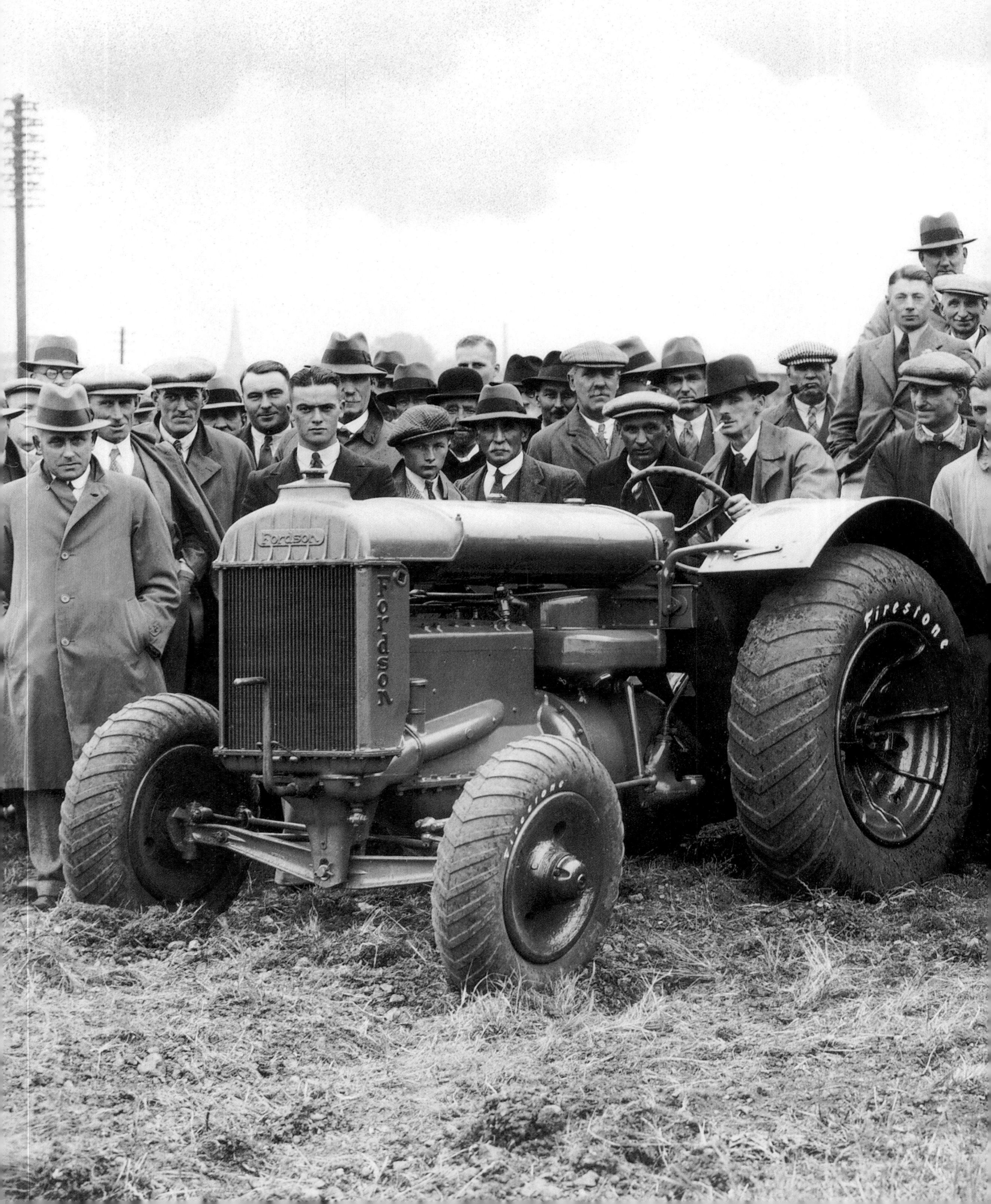
Fordson
Fordson
Firestone

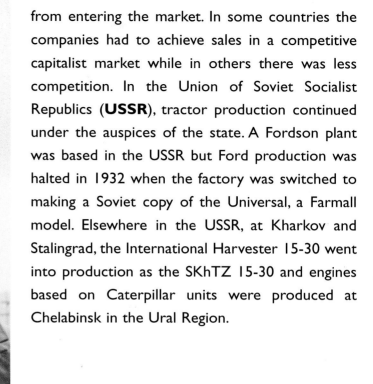

from entering the market. In some countries the companies had to achieve sales in a competitive capitalist market while in others there was less competition. In the Union of Soviet Socialist Republics (**USSR**), tractor production continued under the auspices of the state. A Fordson plant was based in the USSR but Ford production was halted in 1932 when the factory was switched to making a Soviet copy of the Universal, a Farmall model. Elsewhere in the USSR, at Kharkov and Stalingrad, the International Harvester 15-30 went into production as the SKhTZ 15-30 and engines based on Caterpillar units were produced at Chelabinsk in the Ural Region.

Left: A demonstration in June 1934 of Firestone's Golf Course tyres fitted to a Fordson tractor. These tyres were designed to give traction on soft surfaces without damaging the surface. The tyres have been made in sizes suitable for both front and rear wheels.

Below: A Fordson tractor powered by an Ailsa Craig diesel engine.

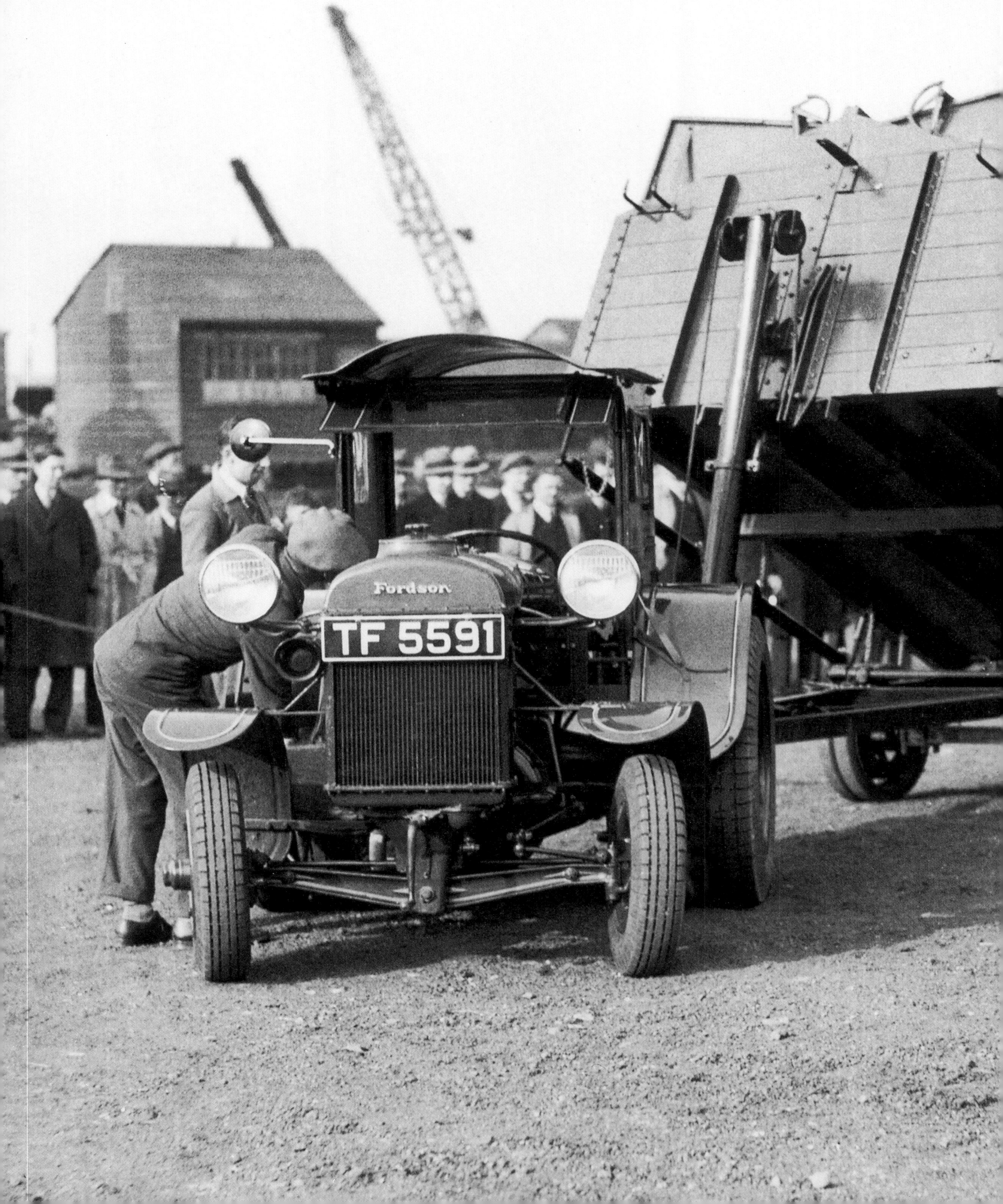
Fordson
TF 5591

Above: A Fordson tractor fitted with a Perkins Leopard diesel engine on display in July 1937.

Left: A Fordson Industrial tractor being demonstrated in November 1931 to trade representatives. The tipping trailer is an Eagle refuse trailer.

Below: A Fordson tractor on display at the Royal Show in England in 1935. The manufacturer claimed the tractor did 'the work of eight horses'.

These pages: The unusual Nectaur tractor relied on its detachable implements for its rear wheels. The drive and steering were connected to the front wheels and a variety of implements could be attached. A general purpose trailer is seen attached (**left**) and disconnected (**above**), In another operation it is seen ploughing (**right**); all photographs are from the pre-war summer of 1939.

These pages: The Fordson
Model N was built at Dagenham,
England, from 1933 onward for
more than a decade. This run crucially lasted through the war
years, when the 'Dig for Victory'
campaign encouraged people to
cultivate all available land to
ensure an increase in
agricultural production despite
shortages created by the war.
The increase in production was
particularly vital as the supply of
American tractors was hindered
by the activity of enemy
submarines attacking shipping in
the Atlantic.

Above: The Nectaur tractor using a two-furrow plough; the differential in the axle is clearly evident.

Left: A Ferguson Brown tractor ploughing with a two-furrow plough in October 1936. Steel lugged wheels aided traction in field conditions such as this.

Above right: The Fendt Dieselross or 'diesel horse'. This one has a sidebar mower fitted.

Right: The Howard DH22 was made in Australia and named after its manufacturer, Cliff Howard.

Next page:. Hall's 1932 25hp Golf Course Fordson paraded at the Masham traction engine rally followed by a 1943 Allis-Chalmers Model B.

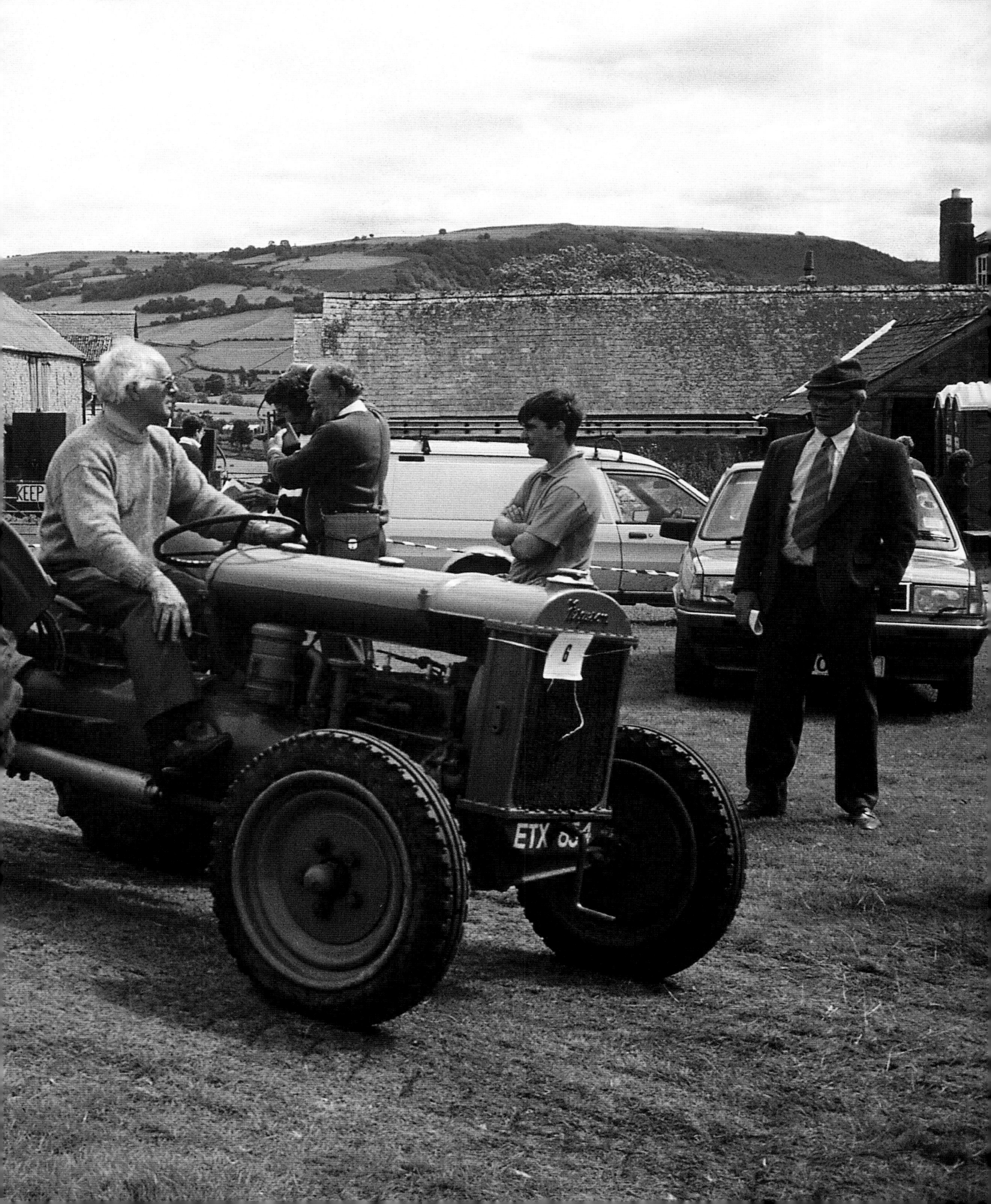
KEEP
6

World Developments

In **Australia** McDonald Imperial Super diesel tractors were in production although the Depression saw the cessation of Ronaldson-Tippet tractor production. In its place came Cliff Howard and the DH22 tractor and saw experimentation with wood burning and gas tractor power.

Above: Société Française Vierzon started tractor manufacture in 1935. This is a 1948 SFV 401 model.

Left: Less than 1,500 Ferguson Brown tractors like this were made in the 1930s through the collaboration of David Brown and Harry Ferguson.

Previous page: Ferguson tractors from different decades; a 1950s' 35 follows a 1930s' Ferguson Brown at an English vintage machinery show.

3 THE 1940s

The shortages of war—such as that of rubber as a result of the Japanese conquests in the Far East—meant that steel wheels came back into use. Prewar designs became standardised and remained in production during the war period with only minor and necessary changes being made. In the postwar years there was considerable activity as tractor makers introduced new products.

North America

In 1941 **Ford** changed the design of the Fordson's grille to a steel item with vertical bars. Many other changes were also made. In fact, by the end of the year Ford had made so many changes, and had ideas for further upgrades, that it changed the designation of

Left: The McCormick Deering standard version of the Farmall M was the W-6. It was also made in orchard form as the OS-6 and O-6 (*seen here*) throughout the 1940s.

the tractor to the Ford 2N. The 2N was similar to the 9N, over 99,000 of which were produced from 1939 to 1942. Over the following five years—1942–47—almost 200,000 2Ns were produced. The 9N's selling price was $585 in 1939. This price included rubber tyres, an electrical system with a starter, generator and battery, and a power take off. Headlights and a rear taillight were an option.

The Ford 2N had a relatively short production run that lasted only between 1942 and 1947. Some of the newest features incorporated into the design of the Ford 2N were an enlarged cooling fan contained within a shroud, a pressurised radiator and eventually sealed-beam headlights. Other changes were made here and there as a result of the constraints imposed by the war. For a while

only steel wheels were available because of the rubber shortage due to Japanese conquests in Southeast Asia and a magneto ignition system used rather than a battery.

When the war ended things reverted to what had been available before. Ford had made 140,000 Model N tractors in England in the war years. The Ford 2N eventually evolved into the Ford 8N, which officially started its production run in 1947

Right: Ford tractor production of the Fordson N (**above**) continued through the war years, while the E27N (**below**) was the company's first new post-war tractor.

Below: All the established American tractor makers added new models to their ranges in the post-war years and John Deere was no exception; producing both more powerful engines and diesel models to its lineup.

and was to last until 1952. 1947 was also the year the much vaunted handshake agreement between Henry Ford and Harry Ferguson was ended regarding the three-point hitch. The two could not come to an agreement when they tried to renegotiate the agreement in the immediate postwar years. Ford decided he would continue using the hitch, but would no longer pay Harry Ferguson for doing so and nor would he call it the 'Ferguson System' any longer. Few official business documents existed and this resulted in a lawsuit which eventually awarded Harry Ferguson approximately $10 million. By the time it was settled, some of Ferguson's patents had expired enabling Ford to continue production of a hydraulic three-point hitch with only minimal changes. Ferguson went on to produce the TE-20 and TO-20 tractors in England and

A tricycle row-crop John Deere Model B tractor. The A and B model designations were superseded in 1952 by numerical model designations.

These pages: Ford's Fordson brand (**above left**) and that of John Deere (**left**) were household names in farming circles by the 1940s and both companies sought to take advantage of developing technology. One example of this is Ford's use of the three-point linkage (**above**).

America respectively which were similar in appearance to the 8N. A completely new line of implements was introduced by Ford. Some of the noticeable differences between the 8N and 2N tractors were the change in wheel nuts from six to eight in the rear wheels, a scripted Ford logo on the fenders and sides of the bonnet and, of course, the absence of the Ferguson System logo which was no longer displayed under the Ford oval even though the tractor still used Ferguson's three-point hitch.

At the outbreak of war the then current tractor and crawler technology was turned to military applications but progressed because of it. The American Army had become interested in crawler and half-tracks as early as in May 1931 when it acquired a Citroen-Kegresse P17 half-track from France for tests and evaluation. US products soon followed; James **Cunningham** and Son produced one in December 1932; in 1933 the Rock Island Arsenal produced an improved model;

Above: Production of the 'grey Ferguson' was accomplished by an agreement between Harry Ferguson and Sir John Black of the Standard Motor Company. This is a 4 x 4 Ferguson converted with a war surplus Jeep front axle.

Right: Ferguson tractors were offered in both diesel and gasoline-engined forms and were manufactured in both the United States and Britain designated TO-20 and TE-20 respectively.

Cunningham built a converted Ford truck later in the year; General Motors became interested; and the Linn Manufacturing Co. from New York produced a half-track. In 1936 Marmon-Herrington produced a half-track converted Ford truck for the US Ordnance Department with a driven front axle. Towards the end of the decade a half-track designated the T7 made its appearance at the Rock Island Arsenal, it was the forerunner of the M2 and M3 models to be subsequently produced by Autocar, Diamond T, **International Harvester** and White. Half-tracks were to provide the basis for a variety of special vehicles as well as armoured personnel carriers, mortar carriers, self-propelled gun mounts and anti-aircraft gun platforms. Throughout the war years the half-track evolved and, although standardised, there are certain differences between the models from the various manufacturers. Vast numbers were supplied under Lend Lease

to the UK, Canada and Russia, and many of the International Harvester produced machines went abroad in this manner. The company also made an amount of 'Essential Use' pickups for civilians who required transport in order to assist the war effort. Civilian production resumed fully in 1946 with the reintroduction of the K-series. Postwar came the KB model and the Travelall, a station wagon-type panel truck.

In 1940 the **Oliver** company introduced a small tractor, the 60, in a row-crop configuration and in 1944 the company acquired crawler maker Cletrac.

By 1940 the **Caterpillar** product line included motor graders, blade graders, elevating graders, terracers and electrical generating sets. By 1942 Caterpillar track-type tractors, motor graders, generators sets and a special engine for the M4 tank were used extensively by the United States in its war effort. The agricultural applications of the larger Caterpillar diesel crawlers such as the D7 and D8 models tended to be reserved for huge acreage farms where multiple implements could be used. The widespread use of Caterpillar products during World War II led Caterpillar to move the emphasis of its operations towards construction in the postwar years.

John Deere's factories produced a wide range of war-related products, ranging from tank transmissions to mobile laundry units. Throughout this period John Deere nonetheless maintained its emphasis on product design, and developed a strong position for the postwar market.

The war also changed the economic position of **Case**. Between 1940 and 1945 three Case plants made in excess of 1.3 million 155mm howitzer shells. Case also made a number of specialised military tractors designated the LA1 model. Alongside these projects, Case made parts for army trucks and amphibious tracked vehicles, and parts for aircraft. The war production did affect Case's sales of tractors to farmers adversely at a time when agricultural production was as crucial as military production, but it did end the effects of the Depression on the company. Despite this, labour relations were not all that they might have been and in 1945 Case employees from the Racine plant went on strike for 440 days, the longest strike in the company's history. This strike had a disastrous effect on Case's dealer base and has been credited by many historians as one of the reasons for John Deere's postwar growth and consolidation of its market. In the aftermath of both the war and the

Previous page: The Farmall Cub was the smallest Farmall tractor. It was built continuously between 1947 and 1979 with only minor upgrades to its specification.

These pages: In the 1930s Harry Ferguson had collaborated with David Brown to manufacture Ferguson Brown tractors and in the 1950s he worked with Sir John Black to produce Ferguson TE-20s (**main picture**). David Brown continued in tractor making without Ferguson (**insets**) and made the VIG/1.

strikes Case looked towards expansion. The company bought plants in Bettendorf, Iowa and Stockton, California as well as the Kilby Steel Company of Anniston, Alabama. With these additional facilities Case sought to produce a wider range of farm machinery including combine harvesters, rakes, ploughs and manure spreaders. The Alabama plant was to be used for the production of new machines including tobacco harvesters for the south east of the USA.

Studebaker was not a tractor maker but a manufacturer of cars and, as for the other US auto makers, World War II interrupted civilian vehicle production as the company's production capacity was turned to the war effort. Studebaker assembled almost 200,000 US6 2.5-ton 6x4 and 6x6 trucks. Half of these went to the USSR as Lend Lease equipment. Studebaker also produced Wright Cyclone Flying Fortress engines and in excess of 15,000 Weasels, a light fully-tracked military vehicle. This latter machine was designed by Studebaker's engineers and can be regarded as one of the pioneers of the light crawler tracked vehicle which has become popular as downsized agricultural machines for numerous specialist applications.

Postwar Developments

All the established American tractor manufacturers quickly added new models to their ranges in the immediate postwar years. **John Deere** was no

exception and replaced the Models H, LA and L with the Model M in 1947. This was followed by two derivatives, the Model MC and MT, a crawler and a tricycle-type, in 1949. The Model M was made between 1947 and 1952 while the MT was made between 1949 and 1952. Production of both models totalled more than 70,000. The MC was John Deere's first designed and constructed crawler machine and in the Nebraska Tractor Tests it achieved a rating of 18–22hp. A variety of track widths were made available to customers and more than six thousand examples had been made by 1952. The MC and industrial variant of the Model M, the MI went on to become the basis of John Deere's Industrial Equipment Division. Another feature of the post-war years was an influx of new manufacturers

New American companies included Brockway, Custom, Earthmaster, Farmaster, Friday, General,

Harris and Laughlin and much further away in Australia the Chamberlain 40 went into production. This machine was built in a war surplus factory and powered by a horizontally opposed 30hp engine. Kelly-Lewis manufactured a basic machine loosely based on a prewar Lanz but the model was soon superseded.

All the other manufacturers made up for lost time by introducing new equipment— **Allis-Chalmers**, for example, introduced the Models G and WD; **International Harvester**

Above right: The Nuffield M4 used a side valve in-line four-cylinder engine which was started on petrol and run on paraffin.

Below right: The engine was a modified version of the unit used in a wartime truck. In the tractor form it produced 42hp at 2,000 rpm.

Below: The prototype Nuffield Model M4 undergoing ploughing trials near Stratford on Avon, England in 1947. The tractor went into production in 1948.

returned to tractor manufacture and replaced the A with the Super A that had a hydraulic rear lift attachment as hydraulics began to be more widely used; and **Massey-Harris** introduced the Model 44.

Massey-Harris brought out a new range in 1947 which included the Model 44 based around the same engine as the pre-war Challenger tractor and fitted with a five-speed transmission but also the Models 11, 20, 30 and 55. The Model 30 featured a five-speed transmission and more than 32,000 were made before 1953 when the company merged with Ferguson. Nebraska Test No. 427 was carried out on the Massey-Harris 44K Standard made by the Massey-Harris Co., Racine, Wisconsin. The dates between which the tractor was tested were September 29 and October 14, 1949. The tractor's specification included a four-cylinder I-head engine, 1,350rpm, 3.875in bore x 5.5in stroke, a six VAuto-Lite electrical system, a Zenith carburettor. The tractor's weight was 5,085lb and the Test H Data obtained was as follows:

o   gear: 3
o   speed: 4.29mph
o   load: 2,608lb
o   slippage: 4.4%
o   rated load: 27.7 drawbar horsepower
o   fuel economy: 10.42hp hr/gal.

Top left: The Nuffield M3, seen in 1948, was a tricycle row-crop version of the M4.

Above left: The M4 featured a PTO, three-point hitch and infinitely adjustable rear wheels.

Left: A Ferguson-equipped with a scraper blade and metal devices on the rear wheels designed to aid traction in difficult going.

Right: This incomplete Nuffield M4 photographed by the manufacturer in 1948 clearly shows the PTO, belt pulley, and sidevalve engine.

116lb of ballast was added to each rear wheel for tests F, G, and H. Test G resulted in a low-gear maximum pull of 4,692lb. Fuel economy at Test C maximum load of 35.66 belt horsepower was 11.3hp hr/gal. Test D rated load of 33.64 belt horsepower yielded 11.19hp hr/gal. Tractor fuel was used for the 45.5 hours of engine running time for this test. The post-war Model 44 was a success for Massey-Harris in the US and 90,000 were made in Racine, Wisconsin. It featured the rounded pre-war styling but was mechanically new. The customer had a choice of four-cylinder engines that used either petrol, paraffin or diesel fuel. A hydraulic lift system for implements was introduced in 1950. The tractor built in largest numbers by Massey-Harris prior to the merger with Ferguson was the Massey-Harris Pony. This was a small tractor and soon proved more popular in overseas markets than in the US and Canada where it was considered too small for many farming applications. It was produced in the Canadian Woodstock plant from 1947.

In 1948 **Oliver** unveiled a new range of tractors to mark its 100th year in business. They were known as Fleetline models 66, 77 and 88 and were identified by a new grille and sheetmetal. Each was fitted with a PTO and a range of engines intended to offer 'something for every farmer'. Engine displacements ranged from the 129cu in (2,113cc) diesel to the 231cu in (3,784cc) diesel in the 99. Other diesel options in the new range included the 77 and 88 models otherwise the tractors used the four-cylinder petrol and paraffin engines. PTO equipment was standard but hydraulic lifts were not.

Ford and Ferguson went their separate ways in 1946 after the latter did not approve of the younger Henry Ford's ideas for the future of tractors. This split led to an amount of litigation and saw Ferguson open his own Detroit factory. The lawsuit which followed suggested that Ford's use of the Ferguson System on 8N tractors was considered a violation of Harry's patent. The final outcome (see above) was that Ferguson won almost

Above: A 1943 photograph of the Women's Timber Corps engaged in forestry work to assist the war effort. A Dagenham-built Fordson pulls the timber trailer.

Above left: The David Brown medium industrial tractor the VIG/1a; it was a slightly improved version of the pre-war VIG/1.

$10 million damages for patent infringement and loss of business from Ford.

In the same period **Allis-Chalmers** introduced the WD-45 model in both petrol and diesel engined forms. The W-45 was a great success and more than 83,000 were made.

Light agricultural products were to prove popular in the post-war years too; as the war drew to a close it became apparent that vehicles such as the Jeep would be invaluable to farmers and ranchers. **Willys** Overland had the foresight to register 'Jeep' as their trademark and began to prepare for the production of civilian Jeeps which were to be tagged CJs. Initially the Jeep CJs were marketed for agricultural purposes by dint of their being equipped with power take offs and agricultural drawbars. They were heavily promoted through a variety of farming tasks such as towing ploughs and disc harrows. The first post-war Jeep was the CJ2A which appeared on the surface to be simply a different colour of Jeep to the military models, but underneath the skin featured revised transmission, axles and differential ratios. More obvious alterations included the inclusion of a hinged tailgate and relocation of the spare wheel to the vehicle's side. There were also numerous detail improvements including bigger headlights and a relocated gas cap. The engine was only slightly upgraded from the MB. Production of the CJ2A lasted until 1949 by which time there had been 214,202 produced. This production run overlapped with the second of the CJs-the CJ3A. This Jeep went into mass production in 1948 and was built until 1953. The main differences between the CJ2A and CJ3A were a further strengthened transmission and transfer case and a one piece windshield.

In the first post-war decade from 1946 to 1954 the **John Deere** company was firmly established as one of the nation's 100 largest manufacturing businesses. Many new products and innovations a were introduced during this period including the company's first self-propelled combine harvester, cotton picker and combine corn head. The latter was one of John Deere's most successful harvesting innovations. The company also made electric starting and lights standard on its post-war tractors including its first diesel tractor. This was produced in 1949 and designated the Model R before the company switched to a numerical designation for its tractors. The Model R was the replacement for the Model D and based on engineering and design that had been started during the war years. It was a powerful machine that gave 45.7 drawbar and 51 brake horsepower in the Nebraska Tests. The large displacement diesel engine that powered the Model R was started by means of an electric start, two-cylinder gasoline engine. More than 21,000 Model Rs were built between the model's introduction and the end of production in 1954.

Europe

At the outbreak of World War II for Britain on September 3, 1939, shortly after the Germans invaded Poland, there were only three major tractor producers in business in the UK—Fordson, Marshall and David Brown. Of these, Fowler's of Leeds was almost immediately engaged on other war work. This meant that a large number of the tractors that would be required in order that farmers could feed the nation through the oncoming war would have to be imported from the USA, initially as ordinarily purchased tractors and later through the lend-lease scheme. This meant that the major American tractor manufacturers would be supplying their products in considerable numbers. Allis-Chalmers, Case, John Deere, Caterpillar, Minneapolis-Moline, Massey-Harris, Oliver, International Harvester and Ford machines were all imported to the UK. Ford also continued production at its UK factory and the machines were

Below: A David Brown tractor driving a thresher on a northern England farm during July 1946.

Above: David Brown made airfield tractors during the war years and reintroduced its pre-war civilian models briefly after the war while it developed a new model.

subtly redesigned to use less metal and repainted to make them less obvious in fields.

The war years saw **Roadless** building a number of types of machine for the British Air Ministry as aircraft tugs and, despite being almost blitzed out of their factory, production was entirely devoted to war requirements, both experimental work and crawler track conversions to imported American tractors including Case, Massey-Harris and Oliver machines.

Ford made 140,000 Model N tractors in England in the war years.

David Brown made a large number of airfield tractors. In the occupied European countries tractor production was mostly halted as the auto industries of these countries were turned over to production for the German occupiers. Later, as the tide of war turned, even the German tractor manufacturers were forced to stop tractor production as a result of bombing and material shortages caused by allied naval blockades. The Germans were desperately short of oil products and farm-

ing was again largely reliant on horses while manufacturers experimented with gas as fuel. Many tractors were converted to run on gas through use of a conversion kit. Hanomag, Deutz, Lanz, Holz, Normag and Fendt all made, or had engines converted to, gas powered tractors. These companies also made numerous wheeled and tracked products for the Wehrmacht.

Post-war Developments

In **Britain** the first new post-war tractor was the Fordson E27N. It was rushed into production at the request of the Ministry of Agriculture. The basis of the machine was an upgraded Fordson N engine with a 3F, 1R gearbox with a conventional clutch and bull wheel and pinion rear axle drive. The production span ran from 1945 to 1951 and in this period various upgrades and options were

Above: The 150,000th Fordson tractor coming off the Dagenham, Essex production line on March 29, 1943. It was considered a milestone in both tractor production and the war effort.

Right and Below right: Details of the implement hitches on David Brown tractors; the top unit uses a hydraulic system.

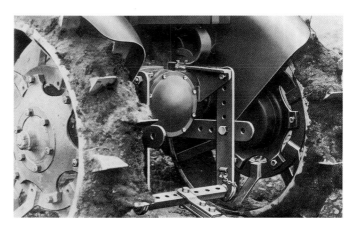

made available. These included electrics, hydraulics and a diesel engined version.

At the same time Roadless converted a large number of the increasingly popular Fordson E27N tractors to half-track machines. The conversion was approved by Ford and received an accolade when it was awarded a silver medal at the 1948 Royal Show which was held in York, England. The Yorkshire, England based David Brown company had made airfield aeroplane tractors during the war years and reintroduced its VAK-1 as the slightly improved VAK-1a until it could introduce the Cropmaster of 1947. This became a popular machine especially in diesel form as introduced in 1949.

William Morris established his reputation by being the driving force behind Morris cars, which became one of the British motor industry's most renowned names. As early as 1926 he had become interested in the tractor market and produced a small crawler tractor based on the track mechanism of a light tank. A dozen Nuffield prototypes were being tested in Lincolnshire by 1946, boosted by the success of these machines plans for full scale production at the Wolseley car factory were made. Nuffield finally entered the tractor market with the Universal which had been unveiled at the 1948 Smithfield Agricultural show. The Universal was offered as the M4 as a four-wheeled tractor and as them as a tricycle type for row-crop work. The engine was derived from the wartime Morris Commercial engine, a four-cylinder sidevalve unit and for the tractor application, started on petrol

and run on paraffin. It produced 42hp at 2,000rpm. Later a diesel variant was offered, initially a Perkins P4 unit then later with a 3.4-litre British Motor Corporation (BMC) diesel following the Austin and Morris merger.

In the post-war years International Harvester was one of the three large American tractor manufacturers to establish factories in Britain, International Harvester opened a factory in Doncaster, South Yorkshire while Allis-Chalmers opted for Hampshire and Massey-Harris, Manchester although they moved to Kilmarnock, Scotland in 1949. International Harvester were the only one to have noteworthy commercial success when they assembled the Farmall M for the British market.

Below: Allis-Chalmers opened a factory in Southampton, Hampshire, England, in the postwar years but moved production soon afterwards to Essendine in Lincolnshire.

Above: Once a common sight on British farms—grey Ferguson tractors ploughing. At times in the early 1950s production exceeded 500 per week.

Above left: The Massey-Harris Model 44 was a post-war success for the Massey-Harris company, and more than 90,000 were made in North America.

Below left: This is the orchard version of the Model 44; the streamlined bodywork allows the tractor to pass trees without damaging growing fruit.

The development of the British Land Rover for farm use closely paralleled that of the American Jeep. Maurice Wilks, the Chief Engineer of the Rover Company bought a war surplus Jeep for use on his Anglesey, Wales estate.

In the early post-war years Rover who had a reputation for building quality motor cars were in a difficult position because of the shortage of steel and the fact that steel was allocated to companies who were producing goods for export which would help to ease Britain's balance of payments situation. Rover had never been a serious exporter of its cars beyond some sales to Britain's colonies. During the war years their Coventry factory had been blitzed and they had moved out to Solihull where they had produced items for the Air Ministry. After seriously investigating the possibilities of producing a small aluminium-bodied car Maurice Wilks and his brother Spencer, also a Rover employee, considered building a small utility vehicle with an aluminium body and four-wheel drive. The intention was that the machine specifically intended for agricultural use would merely be a stopgap until sufficient steel was available for the company to return to building luxury cars. The Wilks brothers delegated much of the design work to Robert Boyle and a number of employees in the drawing office. Rover also purchased two war surplus Jeeps on which to base their design. The designers were given another criterion, too—as much as possible of the Land Rover was to utilise existing Rover components and to avoid expensive tooling costs the panels should be flat or worked by hand. Unlike steel aluminium was not rationed which was another advantage. The prototype Land Rovers had a tractor-like centre steering wheel to

153

Above: World War II changed the face of tractor manufacture. In its aftermath International Harvester (**above**) opened a plant in England as did Allis-Chalmers (**below**), in Doncaster and Southampton respectively.

enable it to be suited for either left or right hand drive markets. Because a conventional chassis would have required expensive tooling engineer Olaf Poppe devised a jig on which four strips of flat steel could be welded together to form a box section chassis.

The first prototype featured a Jeep chassis, an existing Rover car rear axle and springs, a Rover car engine and production saloon gearbox cleverly mated to a two-speed transfer box and a Jeep-like body. It was seen to have potential and an improved version was endorsed and 50 prototypes were built

for further evaluation. A larger and more powerful but still extant Rover car engine was fitted and the centre steering wheel was dropped. In almost this form the 80in wheelbase vehicle was shown to the public at the 1948 Amsterdam, Holland motor show. Orders flowed in especially when early Land Rovers were displayed and marketed at agricultural shows around Britain and the company began looking seriously at export markets. An indication of their potential was perhaps gained when by October 1948 there were Land Rover dealerships in 68 countries. The first production models were different from the pilot batch in a considerable number of ways, some to keep costs down and others to ease production or maintenance. Power take offs and winches were an extra-cost option and between 1948 and 1954 numerous details were refined and improved. Welders, compressors and agricultural equipment were mounted aboard Land Rovers and driven by a centre PTO. An early variant was the Station Wagon which consisted of a then traditional wood framed 'shooting brake' body on a Land Rover chassis. The front wings, radiator grille and bonnet were standard Land Rover panels. While this machine was not the success it might have been the idea was to be recycled more successfully in later years.

Another car company that became involved with tractors was Standard. Following the split with Ford, Harry Ferguson came to an arrangement with Sir John Black of the Standard Motor Company to produce a new tractor in a Standard factory. It was powered by an imported Continental engine and production started in 1946, Standard's engine became available in 1947

and production continued with the new engine, a diesel option was made available in 1951.

The first Ferguson was the TE-20 referred to as the TO-20 in USA. TE was an acronym for Tractor England while TO was one for Tractor Overseas. Both models were not dissimilar to the Fordson 9N but the Ferguson model featured a more powerful engine and a fourth gear ratio. Over 500,000 TE-20s were built from 1946 to 1956 in the UK, while 60,000 TO-20s were built in the USA during 1948-51. This tractor—the TE-20—became nicknamed the 'Grey Fergy', a reference to both its designer and its drab paintwork. It became enormously popular to the extent of being ubiquitous on British farms. J. Wentworth Day writing on farming topics in 1952 said of one farm: 'Three Ferguson Tractors are in use, and this handy, adaptable and powerful tractor seems likely to become a general maid-of-all-work'. Writing at the same time of a Christian community farm he said,

'The Society of Brothers has standardised on one make of tractor—the Ferguson with its range of implements—and find the Ferguson System ideally suited to hill farming. Not only have the tractors handled all the ploughing and other work with cultivators, mower, earth-scoop, Paterson buck rake etc but equally importantly have eased estate management problems by reason of their dependability'. (J. Wentworth Day: *The New Yeomen of England.*; Harrap & Co. London 1952.)

Some Ferguson tractors were put to an unusual task as tractors for the Antarctic Crossing Expedition led by Sir Edmund Hilary. The book entitled The Crossing of Antarctica by Vivian Fuchs and Edmund Hilary contains numerous references to the tractors use on the Commonwealth Trans-Antarctic Expedition. The tractors were used to tow sledges, unload ships and for reconnaissance with the tracked Sno-Cats and Studebaker Weasels also employed. The Fergusons were equipped with rubber crawler-type tracks around the front and rear wheels with an idler wheel positioned between the axles. The expedition was glad of the Fergusons' abilities on numerous occasions and drove one to the South Pole itself! The authors say this about them at one point:

Left: The Allis-Chalmers Model U (**above**) and Fordson Model N (**below**) were built to a similar design and specification by the late 1930s, by which time the generic 'tractor design' had become proven.

Right: A gasoline-engined Caterpillar crawler. Caterpillar specialised in crawlers while some wheeled-tractor makers also offered crawler versions of their tractors.

Left and Right: The McCormick Deering W-4 was the standard counterpart of the Farmall H announced as new for 1940 and powered by an in-line four-cylinder engine.

'Certainly our Ferguson tractors in their modified and battered condition did not inspire confidence in the casual observer but we still felt that their reliability and ease of maintenance would counter-act to some extent their inability to perform in soft snow like a Sno-Cat or a Weasel'. (Vivian Fuchs and Edmund Hilary: *The Crossing of Antarctica*; Cassell, London 1958.)

Renault was the major force in the post-war **French** tractor industry and managed to build more than 8,500 tractors by 1948. Many of these were the 303E model, although it followed this with the 3042 in 1948.

Italian tractor production in the immediate post-war years was in the main a continuation of the prewar Lanz variants while the country, partially destroyed by fighting, reconstructed itself. Same is an Italian company that was founded in 1942 and currently owns the Lamborghini brand. The Lamborghini company that bears its founder's name is perhaps better known for its production of sports cars rather than tractors.

Ferruccio Lamborghini was born in Renazzo di Cento near Ferrara, Italy on April 28, 1916. His enthusiasm for machinery led him to study mechanical engineering in Bologna, after which, during WWII he served as a mechanic in the Italian army's Central Vehicle Division in Rhodes. Upon his return to Italy after the end of the war, Ferruccio began to purchase surplus military vehicles which he then converted into agricultural machines. Then within three years of the end of the war, the Lamborghini tractor factory was designing and building its own tractors. It is hard to say what made Lamborghini turn his attention from agricultural machinery to luxury sports cars.

In **Germany**, occupied and partitioned by the allies things started to return to normality,. There are records of elementary light tractors being made from damaged war surplus Jeeps. Recycling of this nature also took place in Britain and the Philippines. Established companies like Lanz and Hanomag resumed full production once their factories were put in order while MAN produced some 4x4 tractors. The MAN 325 was in production for two decades until the company shelved tractor production in favour of trucks. Another, the Boehringer Unimog, was later sold to Mercedes-Benz and was the forerunner of the Mercedes Benz Unimog. In Austria Steyr returned to tractor manufacture with the Model 180 in 1948, a two-cylinder diesel and the smaller Model 80 in 1949. New companies were founded including Alpenland, Normag, Primus, Stihl and Faun. Of the latter, Faun went on to make trucks and Stihl chainsaws. Fendt returned to the tractor business as did Deutz and the German International Harvester factory resumed production. This had been badly damaged by bombing but once reconstruction work was complete in 1948 it produced the FG12, a version of the F12 Farmall. A restyled diesel variant followed in 1951.

Developments also occurred in the countries that now comprised the Eastern Bloc. The imposition of communist rule meant that the industries of these nations were completely reorganised even were tractor companies already existed. In this way Skoda, Zetor, HSCS, IAR, Ursus and Aktivist came into being or were reorganised in Czechoslovakia, Hungary, Romania, Poland and the DDR respectively. The Zetor tractor company for example, was established in 1946 in Brno, Czech Republic. In the USSR the Universal went back into production and the new Stalinetz 80 was a communist designed crawler tractor.

Below: Case styled its tractors in order to move with the times. The Model L became the LA when fitted with curved sheet metal panels.

Top: A restored Fordson tractor of the type widely employed in Britain during the war years.

Above: A restored International Harvester McCormick W-9 Standard. Belt pulley was standard equipment.

International Harvester used several tradenames as a result of the McCormick and Deering merger. These included McCormick-Deering (**far left**), McCormick (**left**) and Farmall.

Above: 4x4 Land Rovers were used to drive agricultural machinery as this Series I model with a circular saw illustrates.

Next page: The John Deere Model B (**inset**) was the most popular tractor in John Deere. It was one of many tricycle row-crop machines—including this earlier Model G.

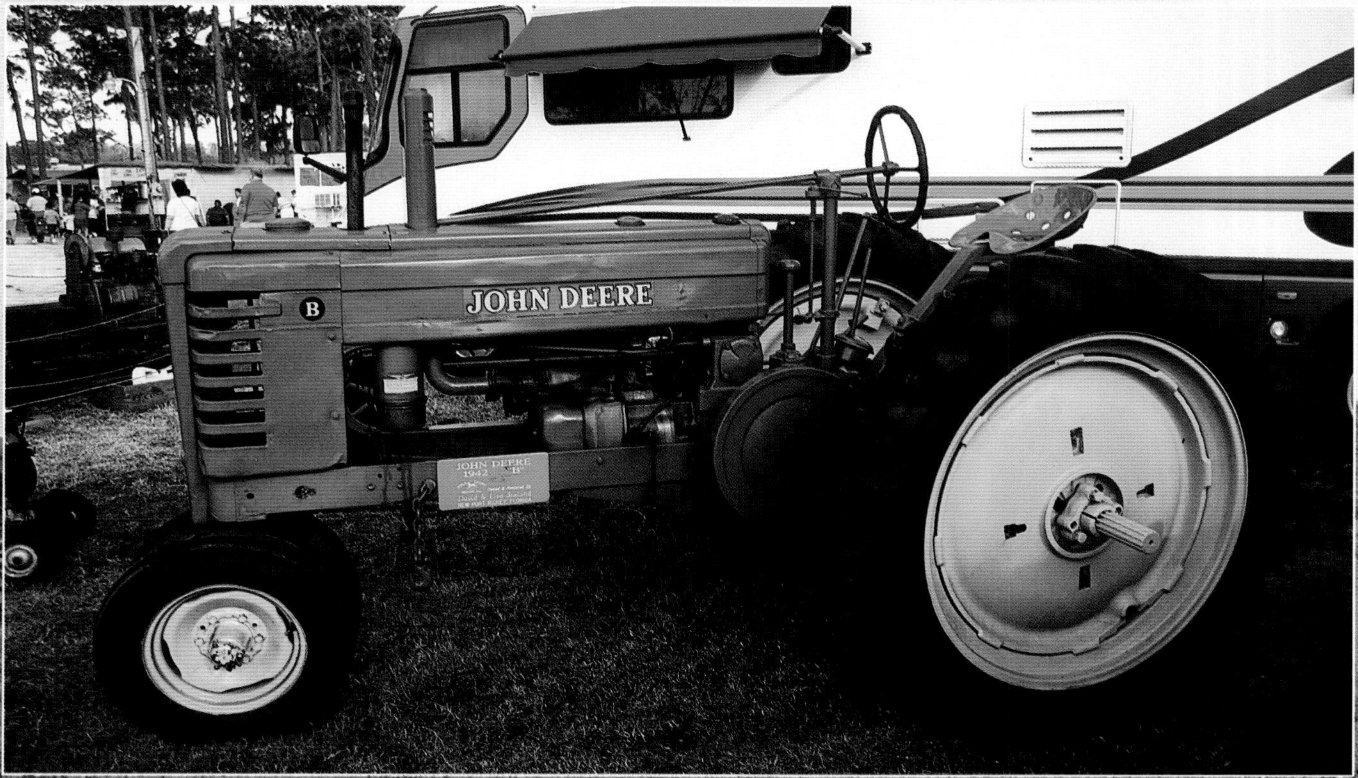
JOHN DEERE
B

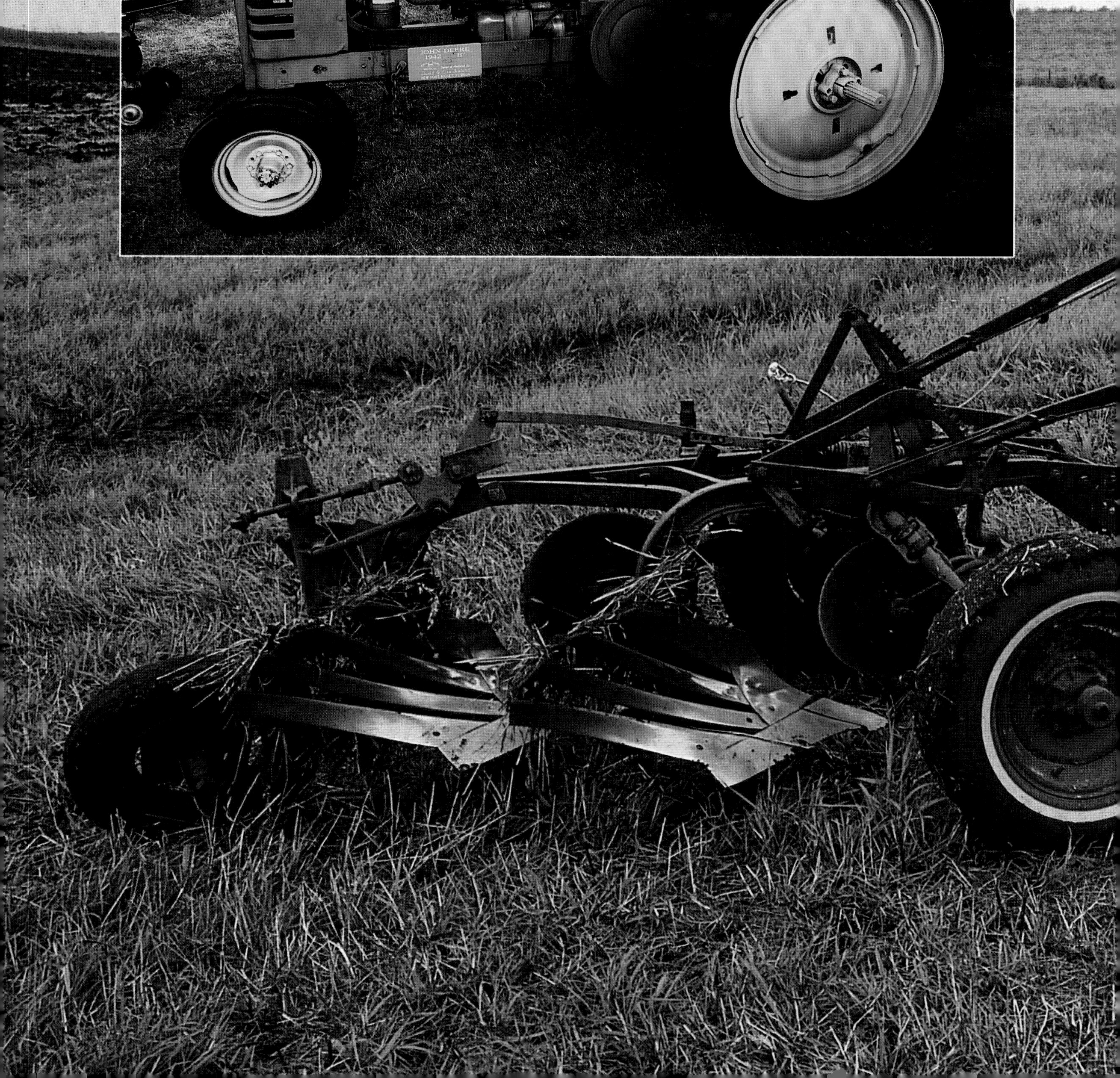

The McCormick Farmall A (**above left**) was one of many compact tractors manufactured by International Harvester. Other makers produced similar machines too (**left**).

Above right: A John Deere 420 equipped with a planter.

Below right: A row-crop International Harvester Farmall M in South Dakota.

Next page: The Centaur was a compact tractor. This is a 1949 Centaur KV model.

CENTAUR
1949 "KV"
S/N 493665
Owned &
Restored By
Ron and Chuck Gaus
PORT CHARLOTTE, FLORIDA

Left: A preserved tricycle row crop John Deere in a farm museum.

Centre left: The three-point hitch is clearly visible on this preserved tractor seen at a vintage rally

Below left: David Brown resumed agricultural tractor production in the years immeditaely after World War II. The Yorkshire-made tractors had a distinctive cowl in front of the driver.

Above right: In Germany, Fendt also resumed tractor manufacture in the years after World War II—this is a post-war Dieselross.

Below right: Hare's restored 1955 International Harvester 50hp Super BWD 6.

Preserved pre- and post-war Fordsons (**top** and **above** respectively) at the Masham, England, Vintage Rally.

Right: The Fendt Dieselross was ideally suited to smaller farms because of its diminutive size.

Left and Right: A Perkins diesel engined Grey Ferguson TED-20. This one which is owned by Michael Thorne and is part of the Coldridge Collection. It is equipped with an air compressor and an air-operated hedge strimmer. The strimmer is seen here as it would be transported.

Tractor makers always prominently displayed their trade names on the bodywork of the machines they made.
David Brown manufactured tractors in Meltham, England, and displayed his name as the brand name (**above**); multinational International Harvester used several brand names including International (**right**). British Marshall coined the name Field Marshall for its post-war tractors (**below right**).

TRACTORS

4 THE 1950s

The 1950s was a decade in which tractor manufacturers prospered; a decade that brought both a booming economy and a period of relative stability to bear on agriculture. Tractor manufacturers had proliferated in the post-war years to the extent that there were 45 manufacturers in the USA, 20 in Britain and approximately 60 in Germany. Diesel tractors became more popular in the USA and the three-point lift became the norm across the industry. The respected Nebraska tractor tests were revived towards the end of the decade although the practical part of the tests now focused on PTO applications rather than belt drives. Massey-Harris and Ferguson combined in 1953 to form what became the Massey-Ferguson Company in 1958.

Left: The 1950s saw continuing development of tractors as manufacturers around the world launched more new models and refined existing ones.

North America

In 1953 the CJ-3B Jeep Universal was introduced and it would stay in production until the 1960s, a total of 155,494 were constructed. In 1953 a CJ-3B with a PTO undertook the Nebraska Tractor Test. It was the subject of Nebraska Test No. 502 which took place between August 28 and September 4. As with all Nebraska tests the speed, torque and fuel consumption were measured. For the drawbar the maximum horsepower recorded was 25.4 and

the maximum for the belt was 35.23. The CJ-5 Jeep was introduced in 1955 and so overlapped with CJ-3B production for a number of years. In many ways the CJ-5 was the last Jeep intended for agricultural use as, gradually, Jeeps became more refined and became sold more for transportation and recreation than as specific farm vehicles.

In the same year Harry Ferguson merged his company with the Massey-Harris company. As the deal was being finalised, reportedly, there arose a

Above: A streamlined row-crop Oliver 70 tractor photographed in Florida.

Right: John Deere switched to numerical model designations in the 1950s and in the same decade began to use multi-cylinder engines in place of the two-cylinder ones that Deere was noted for.

Next page: L. Whitehead's 1961 Massey-Ferguson 35. Production of the 35 models started in the 1950s, but the red and grey colour was not adopted until 1957 after the Massey, Harris and Ferguson merger.

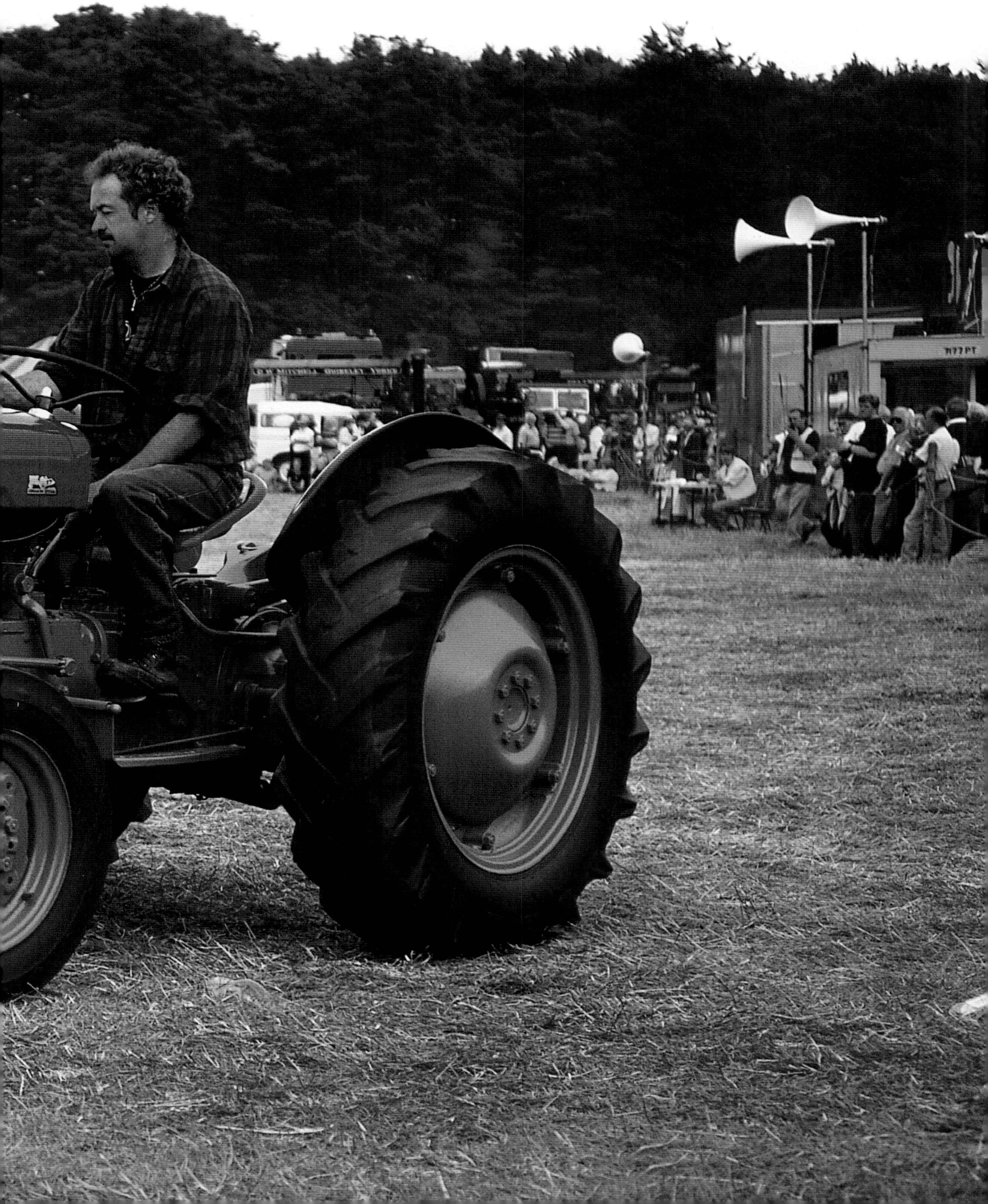

Above: Ford and Ferguson went separate ways after World War II although the two companies' tractors still resembled each other as litigation dragged on.

Right: An International Harvester B-250 ploughing in the English Cotswolds decades after it was made.

question as to exchange rates to be used. Harry suggested a coin toss to settle the matter. He lost the toss, and about a million dollars, but reputedly appeared to not care, no doubt realising that his patents and equipment were in capable hands. For a while this newly formed **Massey-Harris-Ferguson** company produced two separate lines of tractors. They continued with the Massey-Harris line and the Ferguson line as both had loyal followers in the form of dealers and customers.

Nebraska Test No. 603 was carried out in 1956 on a Massey-Harris 333 made by Massey-Harris-Ferguson Inc., Racine, Wisconsin. It was tested between October 25 and November 3, 1956. The tractor's equipment included a four-cylinder I-head engine, 1,500rpm, 3 11/16in bore x 4.875in stroke, 208cu in displacement, 12-38 rear tyres, 6.50-16

front tires with a tractor weight of 5,920lb. The Test H Data recorded was as follows—

o   gear: 3
o   high range, speed: 4.93mph
o   load: 2,262lb
o   slippage: 3.76%
o   rated load: 29.71 drawbar horsepower
o   fuel economy: 10.69hp hr/gal

1,028lb of ballast was added to each rear wheel for tests F, G and H. Test G resulted in a low-gear maximum pull of 5,407lb at 1.31mph with a slippage of 12.08%. Fuel economy at Test C maximum load of

Left: A Ferguson 35 taking part in a vintage ploughing competition.

Below: A restored later Massey-Ferguson 35.

Next page: A restored Massey-Ferguson 35X during a break from ploughing. The rollbar is a later addition designed to protect the operator.

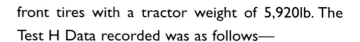

Right and Below: The IH W-9 Standard was a 50 drawbar hp tractor introduced in the 1940s. It stayed in production until 1953 by which time in excess of 67,000 had been made.

39.84 belt horsepower was 12.37hp hr/gal. Test D rated load of 37.11 belt horsepower yielded 12.33hp hr/gal.

The Harris name was eventually dropped by Massey-Harris-Ferguson Inc. and the Massey-Ferguson Co., produced the MF35 in November 1957—its first 'Red and Grey' tractor, the old red of Massey-Harris with the long-running grey of Ferguson.

The Massey-Harris Co. had been producing tractors and competing with both Ford and Ferguson. Once the purchase of Ferguson was complete, the company offered the MH50 and Ferguson 40. These had different bodywork but were mechanically identical and both of these models were based on the Ferguson 35 acquired by Massey-Harris as part of the buy out of Ferguson. In 1958 another new line was introduced. The Massey-Ferguson 65 and 35 were the new models and they were Perkins diesel-powered. Massey-Ferguson acquired Perkins, the diesel engine manufacturer, in 1959. Harry Ferguson established Harry Ferguson Research Ltd. and experimented with numerous innovations including a four-wheel drive system for high performance

sports cars. A modified version of this did later make it into production in the Jensen FF Interceptor.

Ford's 50th Anniversary was in 1953 and that year the Model NAA Jubilee was introduced. This was Ford's first overhead valve engined tractor and had a displacement of 134cu in (2,195cc), producing 31hp. In 1958 Ford introduced the 600 and 800 Series tractors powered by American-made diesel engines although gasoline and LPG versions were also offered. These were followed by ranges of tractors known as Powermaster and Workmaster.

The **Oliver** line up of tractors during the 1950s was the 'Super Series' of 44, 55, 66, 77, 88 and 99 models. The Super 44 was the smallest model in the range and powered by a four-cylinder Continental L-head engine which was offset to permit the operator better visibility for field use. The Super 55 was a compact utility tractor which

on gasoline produced 29.6 drawbar horsepower and 34.39 belt horsepower. The Super 99 of 1954 was available with a three-cylinder two-stroke supercharged GM diesel-powered tractor with a displacement of 230cu in or an Oliver diesel. It was equipped with a three-point hitch and a cab was an extra cost option. The Oliver 66 and 77 row-crop models were perennially popular in Oliver's line up from the late 1940s to early 1960s. Production of

Below: The IH McCormick W-9 standard featured a belt pulley. This one also has rear wheel weights fitted.

Right, above and below: Ford celebrated its 50th anniversary in 1953 and produced the Ford NAA tractor in that year.

Next page: A Massey-Harris-Ferguson Pony of the early 1950s—following the merger of the two companies in 1953 the names of both were used. Later the Harris was dropped to read simply Massey-Ferguson.

Above and Right: A hard worked Nuffield tractor with rollbar protection for the driver.

the Oliver 55 was to continue after Oliver was taken over by White.

Cockshutt manufactured tractors such as the Models 20 and 40. The 20 was introduced in 1952 and used a Continental L-head 140cu in (2,294cc) four-cylinder engine and a four-speed transmission. The Model 40 was a six-cylinder six-speed tractor.

The **International Harvester** W6 was improved in the first years of the 1950s when the tractors were redesigned as the Super W6 and Super M from International Harvester and Farmall respectively. The bore was increased slightly to give the tractors an extra 10hp. In the 1950s International Harvester also made the L, R and S series trucks. In 1954 the Hydraulic Engineering

Company launched its first large machine, an innovative self-propelled swather.

The first **Steiger** tractor was built in North Dakota in the late 1950s. Named after the brothers who designed it, the first Steiger tractor is

Above and Right: The Minneapolis-Moline Model R was introduced in the late 1930s and had a production run that incorporated restyling as well as upgrades such as disc brakes, that meant the machine lasted into the 1950s.

reputed to have gone on to achieve more than 10,000 hours of field time and is now on display in the museum at Bonanzaville in West Fargo, North Dakota. The first Steiger manufacturing plant was a dairy barn on the Steiger brother s farm near Thief River Falls in Minnesota.

John Deere switched to a system of numerical designations for its tractors in 1952 and the tractors that followed were the 20 Series of the mid-1950s, the 30 Series introduced in 1958 and the 40, 50, 60, 70 and 80 series that followed through the 1950s and 1960s. The 50 Series replaced the Model B in 1952, for example, while the 80

replaced the Model R. The John Deere 20 Series tractors were announced in 1956 as a range of six different machines designated sequentially from 320 to 820. Of these the 820 was the largest and was the only diesel model. Despite this, it had a horizontal twin-cylinder engine—John Deere's established configuration. The 820, in many ways an upgraded Model R tractor, produced 64.3hp at 1,125rpm meaning that it was the most powerful

John Deere had built up until that time. The engine displaced 471.5cu in (7,726cc) through a bore and stroke of 6.125in x 8in. Starting was by means of a small capacity V4 petrol engine. The 820 was fitted with a hydraulic lift as standard although the PTO was an optional extra. The 820 weighed in excess of four tons and was capable of pulling six 14-inch furrows. This tractor had a short production run which ended in 1958 but was one of the first in the trend towards bigger and more powerful tractors.

The 30 Series tractors introduced in 1958 were the last two-cylinder machines to be made by John Deere. The company introduced its first four-wheel drive tractor—the 8010—in 1959 and then for 1960 the company introduced a completely new line of multi-cylinder engines capable of meeting the growing demands for more powerful tractors. The new tractors were exhibited in Dallas, Texas in August 1959. There

were four models 1010, 2010, 3010 and 4010 and they were completely new. The smaller models were powered by in-line four-cylinder engines and the larger ones by an in-line six. The famous two-cylinder 'Johnny Poppers' had gone the way of the horse. The six-cylinder 4010 and four-cylinder 3010 were available as diesels although both gasoline and LPG versions were available. The tractors were designed to have higher operating speeds and a better power to weight ratio to increase productivity. The 4010 produced more than 70 drawbarhp while the smaller 3010 achieved more than 50hp. Power steering, power

Below: A 1950s Massey-Ferguson 35 ploughing during the late 1990s.

Above right: A Ferguson tractor with heavily laden trailer undergoing testing in October 1956.

Below right: A David Brown tractor ploughing near Huddersfield, not far from where it was made in Meltham, England.

brakes and power implement raising were up to the minute features and an eight speed transmission enhanced the versatility of the machines. The smaller models in the range, the 1010l and 2010, had features such as adjustable track to make them suitable for row-crop work. The tractors were immediately greeted with acclaim and sales

Above left: A diesel-engined Ford tractor with a rudimentary cab photographed in Lincolnshire, England.

Below left: December 1959, a tractor with the experimental Dowty 'Dowmatic' transmission being demonstrated to the press on a 45°gradient.

Below: A Marshall MP6 tractor on display at an English agricultural show in December 1955.

success. The new range accounted for a significant increase in John Deere's market share over a five year period.

Allis-Chalmers had introduced the WD-45 model with a gasoline engine but in 1955 a diesel variant was offered and marked something of a landmark for Allis-Chalmers by being the first wheeled diesel tractor the company had made. The company had, however, been making diesel engined crawlers since taking over Monarch in the late 1920s. The diesel engine selected for the WD-45 was an in-line six of 230cu in (3,680cc) displacement. A LPG engine option was also made available for the WD-45 and it was the first Allis-Chalmers offered with factory power steering. The WD-45 was a

Above: The post-war version of the Deutz tractor in 'as found' condition.

Left: A British-built International Harvester McCormick diesel-powered B-275 tractor in factory fresh condition in December 1958.

great success overall and more than 83,000 were eventually made.

In 1955 Allis-Chalmers bought the Gleaner Harvester Corporation. The company introduced its D Series in the latter part of this decade. This was a comprehensive range of tractors that included numerous D-prefixed models including the D10, D12, D14, D15, D17, D19 and D21 machines. There were up to 50 variations in the series once the various fuel options of gasoline, diesel and LPG were considered. The D17 was powered by a 262cu in (4,192cc) engine that produced 46 drawbar and 51 belt horsepower in the respected Nebraska tests. There were 62,540 D17s made during the model's ten-year production run. The I.H. Farmall name endured for several decades and tractors such as the Farmall M are fondly remembered.

Case suffered a slump in profits for the years 1950–53 largely because many of the contemporary innovations for tractors including the three-

Above: The International Harvester McCormick diesel B-250 on display at a British agricultural show in December 1955.

Above right: An enthusiast's beautifully restored 'Grey Fergy' in England.

Below right: A 24hp 1950 Ferguson TEA Standard tractor with a 1953 Dinkum digger mounted. In addition to this back hoe the tractor has been equipped with dual rear wheels.

point hitch, were not offered on Case machines, while they were by rival manufacturers. An example of this is in Case's baling machines; in 1941 the company dominated the hay baler market because its baler was the first that picked up the hay as it was towed along. Just before the outbreak of war New Holland had introduced a baler that used twine instead of wire for baling, and more importantly did not require two extra men to tie bales as the Case machine. Case made no improvements to their balers and sales slumped to the extent that Case achieved only 5% of baler sales in 1953.

Case was in the doldrums and the situation was exacerbated by Clausen's relinquishing the presidency in 1948 and being succeeded by Theodore Johnson. Johnson resigned in 1953 to be replaced by John T. Brown. This shift along with impetus from the company's underwriter and a more dynamic board of directors began to turn the company's fortunes around. New engineering practices and an acceptance of diesel engines as well as a new lines of implements including one-way disc ploughs, cotton strippers, disc harrows, Lister press drills and front loaders all took the company forward.

Most important among these advances was the 500 Series tractor. This new line was soon followed by the 400 and, then the, 300 Series tractors. These ranges of tractors were effectively the first completely redesigned Case models since the 1930s. The 500 models of 1953 were powered by an in-line six-cylinder diesel engine that developed sufficient

horsepower to pull five plough bottoms. The 500 also featured power steering, a push-button starter and live hydraulics making them easy to operate.

The 400 Series was unveiled in 1955 and incorporated all of Case's new technology. Power came from an in-line four-cylinder engine with a choice of diesel, gasoline or LPG fuel capable of pulling four ploughs. Final drive was now by means of gears and the transmission had eight forward speeds. A hydraulic vane steering system was fitted as was a three-point hitch and a suspension seat for the driver. The 400 Series was the first TV-

Above left: A 1950 John Deere Model M.

Below left: The 'Economy' tractor was a 14hp compact model.

Below: David Brown unveiled the 14hp 2D tractor at the Smithfield Agricultural Show of 1955. It was of an unconventional design with an air-cooled diesel engine positioned behind the driver's seat

Above: Numerous implements were available for use in connection with tractors as this grey Ferguson equipped with a hinged circular saw indicates. It is seen in the travelling position

Above left: By 1958 David Brown offered a pneumatically operated rear tool bar for the 2D tractor. Around 6,000 2Ds were made over a six year period.

Below left: In September 1957 the Nuffield Universal tractor was made available with a 2.25-litre three-cylinder diesel engine. It also had a PTO and three-point linkage.

advertised models made by the company. Both the 400 and 500 Series models were made at the Racine, Wisconsin plant while the third new model line was to be made at Rock Island and go on sale in 1956. The 300 Series offered customers considerable choice, either diesel or gasoline engines, and transmissions with four, eight or twelve speeds.

The other feature new about the 300 Series was its streamlined styling, which soon became the norm across Case's entire range. The next development in styling came in 1957 and was inspired by the automobile industry when the radiator grille was squared off and inclined forwards at the top giving a slightly concave appearance. This styling would continue through the Case O-Matics and the 30 Series tractors introduced in 1960 and even into the massive 1200 Traction King of 1964.

In 1957 Case merged with the American Tractor Corporation who based in Churubusco, Indiana. This corporation manufactured 'Terratrac'

crawler tractors, and Case continued production of such machines in Indiana until 1961 when it moved production to Burlington, Iowa. One of the results of this merger was the development of the loader backhoe. Case's 320 loader backhoe was the first such machine to be built and marketed as a single unit.

Europe

In August 1951 Harry Ferguson released the TO-30 series, and the TO-35 which was painted beige and metallic green came out in 1954. The Massey-Harris company had chosen the Model 44 tractor to enter the UK market and started by manufacturing them in Manchester, **Great Britain** although operations were later moved north to Kilmarnock, Scotland. This was largely an assembly process because the components were imported from Racine. Models offered included row-crop, high clearance and Roadless-converted half-track models. Around 17,000 were made in total and later approximately 11,000 of a Perkins-engined variant—the 745—were made before production was halted in 1958.

The diminutive Massey-Harris Pony was produced at the Marquette, France factory from 1951 and a decade later production had exceeded 121,000. The first version of the Pony was powered by a four-cylinder Continental engine driving a three-speed transmission with one reverse gear.

Its top speed was 7mph and it produced 11hp. The Pony was, in its initial form, basic but was refined over its production run. Although Canadian production of the Pony was halted in 1954, in 1957 for example, the 820 Pony was offered from the French factory fitted with a five-peed gearbox and a German Hanomag diesel engine. It was further refined for 1959 when it was redesignated the 821 Pony. Around 90,000 of the Pony tractors manufactured were built in the French factory and this was Massey-Harris's first real European success.

Below: The four-cylinder Nuffield Universal Four tractor at the Smithfield show in London in December 1959. The diesel engine displaced 3,403cc (208cu in) and had a compression ratio of 16.5:1.

Right: The Field Marshall was made in a number of variants by Marshall and sons of Gainsborough, Lincolnshire, England.

Another lightweight tractor of the time was the Model 2D unveiled by David Brown at the 1955 Smithfield, London, England agricultural show. It was a small tractor designed for small scale farming applications and row-crop work. Limited numbers of approximately 2,000 were manufactured during a six-year production run. This was a small number for a major manufacturer but accounted for a significant percentage of the market at which it was aimed. Unconventionally the air-cooled two-cylinder diesel engine was positioned behind the driver, and the implements were operated from a mid-mounted tool carrier. Another tool bar was mounted in the traditional rear position and both were operated by a compressed air system. The air was contained within the tubular frame of the tractor. The styling of the other David Brown tractors remained closely based on the rounded VAK

series until 1956 when the new 900 tractor was unveiled. The 900 offered the customer a choice of four engines, although its production run lasted only until 1958 when the 950 was announced. The 950 came in both diesel and petrol forms both rated at 42.5hp. The specification of the 950 was upgraded during the four-year production run with improvements such as an automatic depth control device being added to the hydraulics and a dual speed PTO being introduced. A smaller range, the 850 models were also available at this time.

Left: A trainload of David Brown tractors being despatched from the Meltham, England, works. The 1950s were noted as being a period of 'export or die' for British industry.

Below: A Fordson tractor being serviced during an endurance test. The tractor was worked for 144 hours without stopping in a 1958 test.

Above: A preserved row-crop Oliver 70 tractor at the Florida Flywheelers' Rally.

Left: A row-crop Fordson tractor in factory condition as finished in December 1954.

Next page: This Roadless converted E27N with a high clearance front axle was developed in 1951 for locust control in Tanganyika (now Tanzania). It was fitted with 11 x 36in tyres all around for greater mobility in wet and muddy areas. A Taskers trailer with similar wheels and tyres was towed behind.

Allis-Chalmers had produced its Model B tractor at a plant in Southampton, England from 1948 although by then it was already considered an old-fashioned tractor. The English operation was moved to a plant in Essendine, Lincolnshire soon afterwards and the D270 tractor went into production. In several ways it can be considered an updated Model B, featuring high ground clearance which made it suitable for use with mid-mounted implements and a choice of three engines, all from later Model B tractors. These included petrol and paraffin fuelled versions of the in-line four-cylinder overhead valve unit and a Perkins P3 diesel. The petrol and paraffin engines produced 27hp and 22hp at 1,650rpm respectively. The D272 was a further upgraded version offered from 1959 and the ED40 was another new model introduced in 1960. Disappointing sales of these models were among the factors that ultimately caused Allis-Chalmers to stop making tractors in England, production ceased in 1968.

In 1950 the Caterpillar Tractor Co. Ltd. was established in Great Britain, it was the first of many overseas operations created to help manage foreign exchange shortages, tariffs, import controls and better serve customers around the world. In 1953 the company created a separate sales and marketing division just for engine customers. Since then, the Engine Division has become important in the diesel engine market and accounts for one

Above: Harry Ferguson had developed the three point linkage and the TE20—the famous 'Grey Fergy'—and sold his business to Massey-Harris in 1953.

Right: The earlier Field Marshall tractors were finished in green, while the later ones were painted orange.

Next page: Nuffield were part of the British Motor Corporation (BMC) which became British Leyland, so Nuffield tractors were later badged as Leylands.

quarter of the company's total sales albeit not wholly in agricultural applications.

Through the 1950s the established English Roadless company diversified; it turned its attention to both full track and four-wheel drive tractors. In the same way as the company's half-tracks were built by converting wheeled tractors, so, too, were the full track models. The Fordson Model E was converted and later the Fordson Diesel Major was converted and designated the Roadless J17. At the same time the company also developed a row-crop tricycle conversion for the Fordson E27N

Major for sale in the United States. This stayed in production until 1964. The first Roadless 4x4 tractor appeared in 1956, project-managed by Philip Johnson who was now running the company. He travelled widely to see his company's products in use. While in Italy to visit the Landini company he met Dr Segre-Amar who had founded the Selene company based near Turin. Selene converted Fordson tractors to 4x4 configuration using a transfer box and war surplus GMC 6x6 truck front axles. A working relationship between the two men developed which ultimately led to the Roadless company producing the 4x4 Fordsons in England. This tractor was a great success and subsequently Roadless converted the Power Major of 1958 and the later Super Major in larger numbers and a small batch of Fordson Dextas. International Harvester also marketed a Roadless-

Above: Allgaier were taken over by Porsche in the 1950s and it then offered the Porsche Allgaier tractor powered by a three-cylinder diesel engine.

Left: One of the Ferguson tractors used on the South Pole Expedition by Sir Edmund Hillary. The tractors were equipped with a full track conversion around the front and rear wheels of the tractor and had an idler wheel fitted but were apart from the cab standard. The expedition found the tractors invaluable and praised them highly.

Above right: The production of the Lanz Bulldog in Mannheim, Germany was resumed after World War II, despite the fact that Allied bombing had razed the factory to the ground.

Below right: The BMB President was a small British tractor that appeared in the post-war years. Production and sales were limited because of competition from Ferguson.

Left: The post-war Lanz tractors continued to use diesel engines.

converted 4x4 tractor, known as the B-450 which was constructed in International's Doncaster, England factory until 1970.

The makers of another four-wheel drive agricultural product developed their machines during the 1950s; to make the Land Rover vehicle more capable it was redesigned for 1954, the wheelbase was increased by 6in to make it 86in and the rear overhang increased by 3in. This enabled the rear load area to be increased. A long wheelbase variant, the 107in was also made available as a pickup and these models had a 121.8cu in (1,997cc) engine which had been available in the latter 80in models. Changes were again made to the Land Rover for 1956 when both models were stretched a further two inches to give wheelbases of 88in and 109in. The diesel engine was introduced in June 1957 and by 1958 Rover had produced in excess of 200,000 Land Rovers. In April 1958 the company introduced what was termed the Series II Land Rover. The Series II featured a redesigned body that was wider than its predecessors and featured other minor improvements. In 1957 Nuffield announced the Universal 3, a three-cylinder diesel-engined tractor.

Selene was not the only active tractor maker in **Italy** during the 1950s; one of the Same company's earliest tractors was the 4R20M of 1950 which was powered by a 2hp twin-cylinder gasoline and kerosene engine.

The Landini company continued to produce semi-diesel engined tractors until 1957 as a result

Above and Left: Smaller-sized 1950s' tractors such as the Ferguson 35 (**above**) and Massey-Harris Pony proved popular in Europe and ideally suited to small size farms—this led to large sales.

of this engine type's popularity in many European countries including Germany. For a period Landini and other Italian makers, such as Deganello and Orsi, produced Lanz machines under licence. Landini both redesigned its range of tractors and expanded the company's output; the L25 was announced, this was a 262.3cu in (4,300cc) displacement semi-diesel engined tractor that produced 25hp. It was made in a new Landini factory in Como, Italy. Production of the 55L model, the most powerful tractor to be equipped with the hot-bulb 'Testa Calda' engine, started in 1955. Another key aspect of this redesign was the replacement of the single-cylinder 'Testa Calda' engine used until then, with English manufactured multi-cylinder Perkins diesel power units. This marked Landini's first step towards becoming an international concern. As the trend towards full

Top: Two generations of Ferguson tractors; a Model 35 in the foreground with a TE-20 behind it.

Above: Ferguson tractors became ubiquitous on British farms and were used for any number of heavy-duty tasks—for example this one powered a circular saw.

Above right: A English-built, French-registered grey Ferguson.

Below right: Some of the French-built Massey-Harris Pony tractors, such as this one, were assembled with a German Hanomag diesel engine.

diesel engines continued, in 1957 Landini entered an agreement with the English company Perkins to produce diesels in Italy under licence. This agreement between Landini and Perkins Engines of Peterborough, England for the use under licence in Italy of the English diesel engines, has endured as they are still used and now fitted across the entire Landini range. In 1959 the C35 model was the first Landini crawler tractor to be manufactured.

Fiat had launched a crawler at the beginning of the decade, the Fiat 55, a crawler tractor powered by a 6,500cc (396cu in), four-cylinder diesel engine that produced 55hp at 1,400rpm. The transmission incorporated a central pair of bevel gears with final drive by means of spur gears. It had five forward speeds and one reverse. An option was the means of steering; lever or steering wheel. The six-volt electrical system functioned without a battery and was reliant on a 90-Watt dynamo.

In the mid-1950s John Deere manufacturing and marketing operations had been expanded into **Germany** (and Mexico), marking the beginning of the company's growth into a major multi-national corporation. Elsewhere in Europe increasingly developed technology was being applied to tractors and farm machinery. During the 1950s Bollinder-Munktell abandoned the hot-bulb engine and introduced the BM35 and BM55 models with direct injection diesel engines. These tractors also featured five-speed transmissions and an optional cab. Bollinder-Munktell later merged with Volvo. In 1952 Claeys launched the first European self-propelled combine harvester. Steyr introduced the models 185 and 280, three and four-cylinder diesel tractors respectively. Ursus, a state-owned Polish tractor maker, made tractvehiclesors were very similar to those from other state-owned companies elsewhere in the eastern bloc including 1950s models from the East German (DDR) ZT concern and Czechoslovakian Zetor. Tractors were also produced in Bulgaria under the Bolgar name.

Right and Below: While the Massey-Ferguson logo (**below**) suggested both the merged companies' names and the famous 'Ferguson system', Porsche acquired Allgaier to produce tractors in Germany

While the Deutz D30 was powered by an air-cooled diesel engine (**left and below right**) there was a Hanomag diesel engine in the French built Massey-Harris Pony (**below left**). French production of the Pony outlasted that in Canada and the tractor competed with imported tractors (**right**).

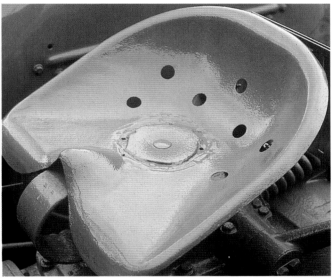

Left: Richard Clarke's restored 1950 Ferguson TEA-20 on a Cheshire, England farm.

Above: The grey Ferguson tractor featured a PTO shaft (**top**) for driving implements, a sprung seat (**centre**) and a basic three-spoked steering wheel (**above**).

233

Previous page: An Allis-Chalmers WD-45. This was one of the company's first new postwar tractors and the diesel variant was the first wheeled diesel tractor made by the company.

Above: Allis-Chalmers manufactured tractors including B, 270 and 272 models in Southampton, England, in the postwar years, but soon moved the plant north to Essendine in Lincolnshire.

Above left and Left: Details of the BMB President tractor; dashboard and steering box. This diminutive British-built tractor was only made in small numbers.

Above right: The styling on John Deeres evolved from one model range to the next, as indicated by the 20 and 10 series tractors.

Below right: The grey Ferguson was austere in terms of fittings and driver comfort.

Top: The John Deere 20 series including this row-crop variant made its debut in 1956.

Above: An Allis-Chalmers with a sidebar mower and tyres suited to mowing golf courses and playing fields.

Right: This four-wheel drive Ferguson is equipped with a rear-mounted shovel reliant on the three-point hitch, rather than hydraulics like later back hoes.

Above: Michael Thorne, proprietor of the Coldridge Collection on his four-wheel drive Ferguson TE-20 in England.

Left: 1950s Allis-Chalmers WD-45 in an American field.

Below: Massey-Harris Pony at a Florida vintage event.

Next page: A row-crop John Deere with an unusual cable-operated rear-mounted shovel.

Left: A John Deere Model 40 in its distinctive green and yellow colours.

Below left: Nuffield's tractors were launched in the post-war years and marketed with a variety of engines as were the Fergusons (**right and bottom right**) seen here with differing farm implements.

Above and Right: A 1951 Nuffield M3 tractor being paraded at the Masham, England steam engine and fair organ rally. The first M3 diesel models were powered by a Perkins P4 engine as the emblem on the grille suggests, although this engine was later superseded by a 3.4-litre BMC engine. This tractor is owned by T. Speight.

Next page: The Farmall M was one of the range of IH tractors that featured Raymond Loewy's redesigned bodywork and introduced in the 1940s. As more power was required by farmers in the 1950s there were numerous variants including the Super M and the diesel, the M-D.

An array of restored tractors at the Florida Flywheelers' event; John Deeres (**left and below**). IH McCormick (**below right**), and a specialist rear-engined unit (**right**).

These pages: An array of restored tractors at the Florida Flywheelers' event; the styling of the row crop John Deere Model A (**above left**), contrasts with the later 420 model (**left**). The same applies to the McCormick Deering O-6, an orchard tractor (**above**) and later IH model (**right**).

Above and Right: The four-wheel drive converted Ferguson tractors (**above**) used World War II surplus Jeep axles (**right**) to provide drive at the front axle. This one is now part of the Coldridge collection.

Left: A Farmall patiently awaits restoration in an English field.

Left: A diesel Field Marshall driving a thresher from its belt pulley.

This page: Güldner were one of numerous German tractor makers manufacturing farm machinery in the 1950s.

Next page: T. Russell's 46hp Chicago, Illinois-built International Harvester T6 TracTracTor. Lend lease tractors such as this gave decades of service on British farms, this one earning its living in Lincolnshire until it was bought for preservation.

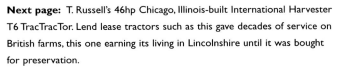

Left: A Nuffield diesel in farm condition using its belt pulley to drive a vintage thresher.

Below left: A restored Ferguson 35 at the Masham vintage rally.

Above right: A similar model Ferguson in farm condition.

Below: A Ferguson TE-20 also in farm condition.

Top: A 1942 International Farmall A, this model was replaced by the Super A in the post-war years.

Right: A diesel-engined Field Marshall on an upland Yorkshire, north England, farm.

Above: The Massey-Harris 820 Pony proved popular in Europe, in particular in France where many were built.

Left: An International Harvester McCormick Standard W-9 equipped with a rear winch.

Next page: A Güldner tractor; Güldner were one of around 60 tractor makers active in Germany during the 1950s.

Top: An Allis-Chalmers awaiting restoration at the Coldridge Collection.

Above: A 'Grey Fergy' still use on a Yorkshire farm in the north of England.

Above left: A well-used Massey-Ferguson 35 on the Island of Gozo near Malta.

Below left: A restored Allgaier tractor made in Germany in the 1950s.

Right: Restored Allis-Chalmers and Ferguson TE-20 at an English vintage event.

Below: The diminutive BMB President was aimed specifically at market gardeners who worked much smaller plots.

This page: Details of the radiator grille badges of the BMB President (**right**) and the Fordson Super Dexta (**below right**) and a 1950s Fordson (**below**).

Far right, above: R. Middleton's Ferguson TE-20 fitted with the P3 conversion, Howard reduction box and half tracks.

Far right, below: A similar conversion except that the full track was carried out to the Ferguson tractors used on the South Pole expedition.

Above: Two examples of restored, British-made Field Marshall tractors. They differ in details such as the type of maker's logo on their enclosed bodywork.

Above: A 1950s grey Ferguson now part of the Coldridge Collection of tractors fitted with crop-spraying equipment and row-crop wheels and tyres.

These pages: A 1951 Turner diesel tractor restored by Michael Thorne of the Coldridge Collection.

These pages: A 1953 Ferguson TET 20 powered by a 26hp diesel engine and now part of the Coldridge Collection. The tractor, which had originally been painted yellow, is restored as a Surrey County Council grounds maintenance tractor. It has lights and mudguards to make it road legal.

These pages: A petrol-engined grey Ferguson, now part of the Coldridge Collection, restored and fitted with a half-track conversion. The crawler tracks require the additional idler wheel fitting within the tractor's wheelbase and were designed to enhance the tractor's performance in muddy conditions.

Pages 280–283: The industrial variant of the Ferguson FE35 tractor, powered by a 34hp diesel engine. This one is part of the Coldridge Collection and was originally used for maintaining a school's playing field. During restoration it was fitted with a forklift attachment. The forklift is capable of lifting one ton but requires front wheel weights and weights on the frame at the front of the tractor in order to do so.

These pages: The Coldridge Collection of tractors includes this proto-type Massey-Ferguson 35 from 1957.

These pages: A French Massey-Ferguson 130 tractor. This deluxe vehicle was powered by a 30hp diesel but was not popular in Britain because the more powerful MF-135 was available at a similar price. This 1957 machine was restored in 1996 and is now part of the Coldridge Collection.

This page: This MF 65 was manufactured in 1959. It is now part of the Coldridge Collection where it has been restored..

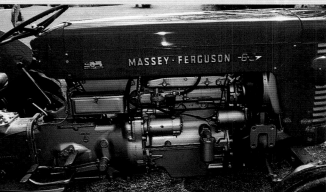

5

THE 1960s & 1970s

60s

The 1960s saw increasing refinements made to the diesel engine as turbochargers and intercoolers were developed. Take-overs and mergers characterised the industry as it adapted to the economics of the decade.

North America

In 1959 the American tractor manufacturers adopted industry standards for the increasingly popular three-point hitches to make farm implements more versatile by increasing their interchangeability between makes. To meet this standard **International Harvester** offered new tractors for 1960—the 404 and 504 models—that boasted the first American-designed draft-sensing three-point hitch. The 504 also came with power steering, while both models featured dry air cleaners and a means of cooling the hydraulic fluid. International Harvester embraced the developing

Left: The 1960s were an exciting period of increasing technology and modern styling for the entire tractor industry.

Above: The three-point hitch devised by Harry Ferguson revolutionised the use of implements with tractors.

Right: The advent of the mini-tractor led to the development of ride-on lawnmowers and compact tractors.

Below right: Tractor cabs offer protection to the driver and have become advanced in design.

Far right: A high clearance version of the 1960s John Deere 3020 diesel.

Next page: A Fordson ploughing in a vintage ploughing match in the English Cotswolds.

technology and developed better transmissions and four-wheel drive systems as farming became altogether more mechanised.

In the same year **Allis-Chalmers** changed the shade of the paint in which its tractors were finished; an even brighter hue, known as Persian Orange Number 2, was selected. This was contrasted on tractors by cream-painted wheels and radiator grilles. The main models to benefit from this change were the D10 and D12 tractors which had

Left and Above: Tractors face decades, not just years, of hard work. Increasingly many are restored and cherished as these two compact John Deeres (**above**).

superseded the Models B and CA. The production run lasted ten years but in all less than 10,000 examples were made.

The **White** Motor Corporation of Cleveland, Ohio became established in the tractor manufacturing industry during the early 1960s when the already established corporation bought up a number of relatively small tractor-producing companies. These companies were established ones with a history of involvement in the tractor industry often going back to its earliest days. In 1960 White bought Oliver but continued to produce the successful and popular Oliver 55 models throughout the decade. Oliver had been in business since 1929 when it, too, had come about as a result of combining several small companies. The Oliver Chilled Plough Works, Hart-Parr, Nicholls and Shepard and the American Seeding Machine Company were combined under the Oliver brand.

White then purchased the Cockshutt Farm Equipment Co. in 1962. Following these substantial acquisitions, White took over Minneapolis-Moline a year later. The brand names of all three companies were retained by White until as late as 1969, when the entire company was restructured as White Farm Equipment. AGCO purchased White Tractors in 1991 and with it acquired the rights to the Minneapolis-Moline name.

As we have seen in the last chapter, the Canadian-based Massey-Ferguson company was formed through the merger of Massey-Harris and Ferguson, as a result of which produced tractors all around the world. The first post-merger tractor was the MF-35 of 1960. This was soon followed by a range of tractors including the MF-50, MF-65 and MF-85.

In 1963 in a move to a related, if smaller, type of machinery **John Deere** entered the lawn and grounds' care business. During the second half of the 1960s the company's industrial equipment line was expanded to include motor graders, four-wheel drive loaders, log skidders, backhoe loaders, forklifts, excavators, new models of elevating

scrapers, utility crawler dozers and loaders were all introduced. In 1969 John Deere entered the insurance business with the formation of the John Deere Insurance Group that also offered credit-related insurance products for John Deere dealers.

In 1963 the Hydraulic Engineering Co. was incorporated as a public company with the name of **Versatile** Manufacturing Ltd. In 1966 Versatile, operating out of Winnipeg, Canada, became involved in the manufacture of huge four-wheel-drive tractors that produced power in excess of 200hp. Later Versatile Farm Equipment Co. became

Below: A Massey-Ferguson 135 with a plough on the three-point hitch.

Above right: A Cuthbertson converted Series II Land Rover 109; this bolt-on crawler conversion was offered in the early 1960s.

Below right: A preserved example of the Cuthbertson Land Rover in Scotland. Crawler tracks are fitted to a subframe.

Next page: An International tractor equipped with a PTO-driven mower, note PTO shaft from tractor to centre of mower.

part of Ford New Holland Americas, N.H. Geotech's North American division.

By the 1960s **Ford** was producing tractors in Brazil, India and England as well as the USA. The Model 8000 was the first Ford tractor to have a 100hp engine, it displaced 401cu in (6,571cc) while the smaller displacement Model 6000 produced 66hp. In 1964 Ford Tractor Operations were moved to Basildon, England while Ford began the manufacture and assembly of Ford tractors in Sao Paulo, Brazil.

A major development of the late 1960s was the change of ownership of **Case**. The Kern County Land Company were a majority shareholder in Case and in 1967 this company was acquired by Tenneco Inc from Texas.

Top: David Brown launched a new tractor in 1961; it was the Model 880 Implematic seen here ploughing near Huddersfield, England.

Above: By the early 1960s the Fordson Major diesel was available with power steering—as seen here.

Above Right: Ford exported its tractors around the globe from the Dagenham works. This batch, photographed in 1965, is being loaded aboard a train for rail transport to the Avonmouth Docks for export to Canada.

Below right: A Ford Dexta—one of the batch being exported to Canada—is craned onto a rail truck for its journey to the docks near Bristol, before being shipped to Vancouver.

Europe

In **Britain** the Universal tractor Model 4/60 was announced in 1961, this was a 60hp tractor and the increased power had been achieved by boring out the existing 3.4-litre engine to a displacement of 3.8 litres. Production was eventually moved to the Bathgate plant in Scotland via the Morris Motors plant at Cowley in Oxfordshire. The Nuffield tractors were given a facelift in 1964 and became the 10-60 and 10-42 models. The models were redesigned for 1967 but before these newest models could find widespread acceptance Nuffield were involved in the merger that created the British Leyland Motor Corporation so making the Nuffield tractors into Leylands.

David Brown offered a 35hp diesel from 1960 onwards. It stayed in production until replaced by the three-cylinder diesel-powered 770 in 1965. A number of Oliver tractors were produced by David Brown in its Meltham, England factory for the Oliver Corporation. This had the dual advantages of increasing the volume of tractors produced in the David Brown factory and allowed Oliver to rationalise its range of models in production in the US without reducing the choice offered to its customers. The David Brown 850 became the Oliver 500 and the David Brown 950 the Oliver 600. The tractors were finished in green and white and changed only in details from their David Brown equivalents. The David Brown 950 came in both diesel and petrol forms and both were rated at 42.5hp. The specification of the 950 was upgraded during the four-year production run with improvements including the addition of an

Below: This Deutz D40 was powered by an air-cooled diesel engine and produced by the German manufacturer in 1961.

Above right: The Track Marshall was a crawler produced by the English Marshall company; this is a 1964 model.

Below right: An experimental fuel cell-powered tractor photographed in Britain during 1960.

Top: Tractor driver's view from a tractor fitted with a typical early cab.

Above: A compact BMB President.

Above left: A 1964 photograph of Ford's 1964 range of tractors; the Dexta, Super Dexta, Major and Super Major, designated Models 2000, 3000, 4000 and 5000 respectively.

Left: The BMC Mini tractor of the mid-1960s was powered by a 948cc in-line four-cylinder diesel engine that produced 15bhp. The tractor was physically small and aimed at small-scale farming applications.

automatic depth control device to the hydraulics and a dual speed PTO being introduced. A smaller range, the 35hp diesel 850 models were also available at this time offered from 1960 onwards. The agreement between Oliver and David Brown lasted around three years and saw in excess of 2,000 tractors exported to the United States.

Around the end of the 1950s, the Hounslow, England-based Roadless company focused its attention on the light 4x4 Land Rover which had now been in production for a decade. The Forestry Commission was experiencing some difficulty with its use of conventional Land Rovers which were prone getting stuck on rutted forest tracks and hampered by fallen trees for cross country use. The Machinery Research Officer for the Forestry Commission, Colonel Shaw suggested that tractor-type 10x28in wheels should be fitted to the Land Rover. A prototype was built and evaluated in the Alice Holt Forest in Hampshire, England. The end result concluded that the machine had potential but required further development in order to be viable. A 109in wheelbase Land Rover was

despatched to the Hounslow premises of Roadless for a redesigned and properly engineered conversion to be effected. Roadless used a combination of components to make the machine functional; they retained the original gear-and transfer boxes and coupled them to a pair of Studebaker axles with GKN-Kirkstall planetary hub reductions and the same 10x28in wheels that had been used before. The front axle had a 14in wider track than the rear in order to facilitate the provision of sufficient steering lock to retain the vehicle's manoeuvrability. The resultant turning circle was approxi-

Left: A restored 1965 BMC Mini tractor owned by B. Manby in the parade of tractors at the Masham, England rally.

Below left: A restored 1960 David Brown Implematic owned by D. Binks at the same rally.

Below: The diesel engine and belt pulley of a restored Fordson Major Diesel.

Next page: This restored 50hp Fordson Major is a 1958 example that was supplied new on oversize tyres but production of this model continued into the 1960s.

mately 40ft. In order to accommodate the large wheels and tyres the normal Land Rover front wings were removed and replaced with huge flat wings while the rear ones were fabricated in the manner of tractor rear wheel arches. With the change in gear ratios and wheel diameter the complete machine was capable of approximately 30mph.

Roadless prepared a second prototype which was sent to the Special Projects Department of the Rover Co Ltd. It was thoroughly tested which led to Roadless making some further modifications to the Land Rover including strengthening the chassis and altering the axle clearances. After a couple of years of tests Rover approved the conversion. This approval meant that the converted Land Rover could be marketed as the Roadless 109 from 1961. In December of the same year the 2,286cc petrol-engined model retailed at £1,558 and the diesel variant £1,658. A special pickup body was listed as an extra at £172. The Roadless

was capable of returning 10mpg and had a 15-gallon tank. It was reported that the machine understeered when cornering at 15mph, the steering was heavy at low speeds and that the machine was capable of wading in up to two and a half feet of water. This latter fact was one of the features that Roadless Traction Ltd mentioned in its adverts along with pointing out information such as the wide track ensured stability on side slopes—a useful asset for forestry work—and so on. The Land Rover project was of minor importance for Roadless, as by the mid-1960s the major portion of its business was the production of four-wheel drive tractors. For example, approximately 3,000 Roadless four-wheel drive Fordson tractors had been manufactured when production of the Fordson Super Major ended in 1964.

Developments occurred in **Italy** during the early years of this decade. In 1960 Landini became part of Massey-Ferguson and in 1962 Fiat entered

Above and Right: The David Brown 880 superseded the 950 model in 1962. This one has a roll bar designed to protect the operator in the event of the tractor rolling over on a hillside.

Below: The Farmall 100 series tractors were announced in the mid-1950s and began a progression of tractors designated by numbers. The 400 was replaced by the 450 and following a redesign later became the 460.

Below right: By 1960 Allis-Chalmers was using a bright orange colour scheme with contrasting cream radiator grille and wheels. The company made less than 10,000 of the small 25.8 drawbar horsepower D10 and D12 models between 1959 and 1968.

a joint venture with the Turkish company KoA Holding in Ankara, Turkey, known as Turk Tractor. By 1966 Fiat had created a Tractor and Earthmoving Machinery Division. The Lamborghini 5C was a crawler tractor unveiled at the Paris, France Agricultural show in 1962. The tractor was unusual in that, as well as crawler tracks, it had three rubber tyred wheels which allowed use on the road by means of lifting the tracks clear of the road surface. The tracks were driven from the rear sprockets and it was to these sprockets that the wheels were attached for driving on road. The

small front wheel was effectively an idler and steering was achieved by moving the same levers as slewed the tractor around when running on tracks. The 5C was powered by a three-cylinder diesel engine that produced 39hp.

The pattern of mergers continued elsewhere; in **France** SFV—Societe Francais Vierzon, a company that had entered the tractor market in 1935 with a machine not unlike the Lanz Bulldog and endured despite World War II—was acquired by Case in 1960.

In 1964 Sperry New Holland purchased a major interest in Claeys, by now one of the largest combine manufacturers in Europe. Sperry New Holland launched the haybine mower-conditioner, Model 460. This was capable of accomplishing what had previously required two or three machines and was perceived as a significant innovation in hay harvesting technology. A Hanomag twin-cylinder diesel engine was offered in the European versions

Left: A mid-1960s Zetor 3011 diesel made in Czechoslovakia. Zetor called its hydraulic hitch system 'Zetormatic'.

Below left: A BMC Mini tractor with a cab; this tractor struggled to compete with secondhand grey Fergusons for sales in the mid-1960s.

Below: There is something timeless about a restored tractor in an English field.

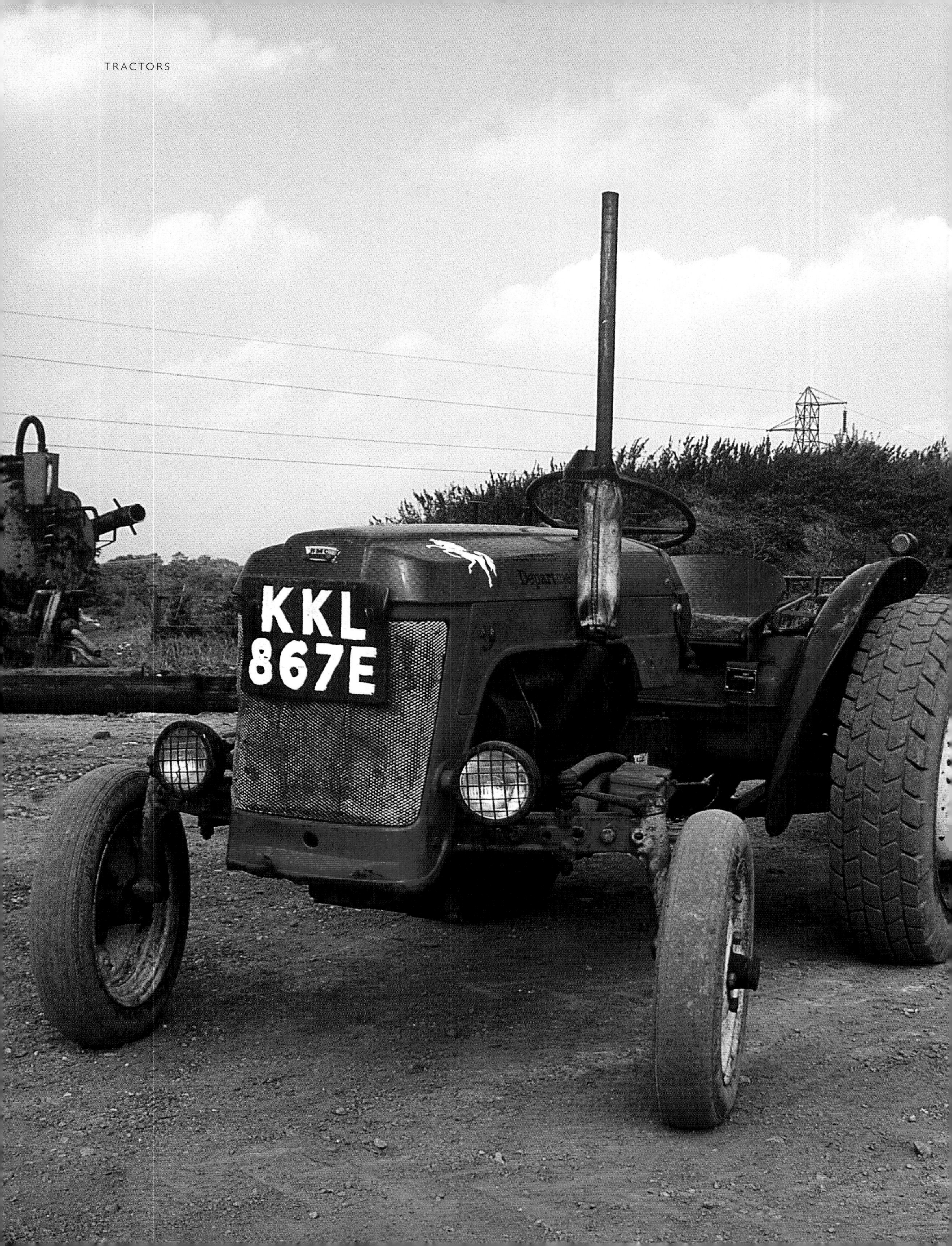
KKL
867E

Above: All the major manufacturers added compact and mini-tractors to their ranges during the 1960s. this is the International Cub Cadet 100.

Left: A 1966 BMC Mini tractor in farm condition—with non-aggressive grounds and lawn type tyres, it hauls a more modern trailer.

of the Massey-Harris Pony as built in France which outlasted Canadian production. Hanomag continued to offer its R-range of tractors including the small R12 and R24 models alongside larger R45 and R55 models through the 1960s.

Throughout its history, **Czechoslovakian** Zetor has endeavoured to pioneer various innovations, Zetor had one of the first hydraulic hitch systems, known as Zetormatic which it introduced in 1960. Later Zetor was the among the first tractor companies to manufacture fully integrated safety cabs with insulated, rubber mounted suspension. In other places behind the Iron Curtain tractor manufacture continued. In 1951, under the auspices of the communist regime one company

Below: The John Deere 5010, seen here with a forage harvester and drawbar trailer, was a large diesel-powered tractor and one of a new range of tractors for the 1960s.

Above: A John Deere combine harvester in the field; air conditioning, computer control and even cabs were still far in the future when this picture was taken in the 1960s.

Above: A John Deere Model 55-H sidehill combine harvester operating in Washington State, USA.

name was changed to Red Star Tractor but from 1960 onwards the products were sold under the DUTRA brand name. This was an amalgamation of the words Dumper and Tractor representing the company's product range. Tractors were exported both to other Eastern Bloc countries and beyond.

Below: This chrome badge on the side of a Ford tractor identifies it as a Fordson Major with a diesel engine.

Below left: A 1960s' David Brown still in use on a farm more than three decades after it was made.

Bottom: A tractor in farm condition; many hours will be required to restore it to the as-new condition of many preserved vintage tractors.

Right: ATCO is a noted English manufacturer of lawn mowers; the company also made these lawn tractors.

Below right: Fordson Major nose and radiator grille badge detail. This badge is on the side of the tractor.

Left: Numerous small agricultural machines were offered by a variety of manufacturers for nursery and small scale farming tasks. This tiller-steered three-wheeler is one such British machine.

Below: Gradually, tractor styling became more angular and modern looking; both mudguards and radiator grilles were distinctly less curved as exemplified by this Massey-Ferguson MF-135.

Right and Below right: Someca was a noted French maker of tractors. This diesel-powered tractor was one of the company's products.

These pages: Roadless Traction Limited produced a four-wheel drive conversion for Fordson tractors using a system patented by Selene. It used a war surplus GMC 6 x 6 truck axle (**right and far right**) which was driven by the driveshaft from a transfer box. Manuel Roadless was the Selene trade name for the conversion. This particular Roadless 4 x 4 Fordson (**below right**) was used as a forestry tractor in Scotland, hence the protective cage for the driver (**below**). It is now owned by Graham Clark.

322

A John Deere 40 combine harvester preserved in South Dakota, USA.

Above left: Renault, Somua and Someca, were among the noted French tractor makers, although Massey-Harris also had a factory in France.

Left: A John Deere 630 tractor. The 30 Series tractors were introduced in 1958 and were followed sequentially by the 40, 50, 60, 70, and 80 series through the 1950s and 1960s.

Above: A John Deere 1010 tractor; this series was powered by multi-cylinder in-line engines and designed to meet demands for more powerful tractors.

Right: A tricycle row-crop Oliver tractor. The White Motor Corporation bought the Oliver company in 1960 but persevered with the brand name.

Next page: A hard-working tractor seen in Florida with a mower. Numerous dings and dents are testament to long hours of use.

Right: A tractor in as-found field condition (**above**) and a restored John Deere with an additional radiator grille guard (**below**).

Below: A John Deere 1010 tractor; these 1010 series models were powered by in-line four-cylinder engines.

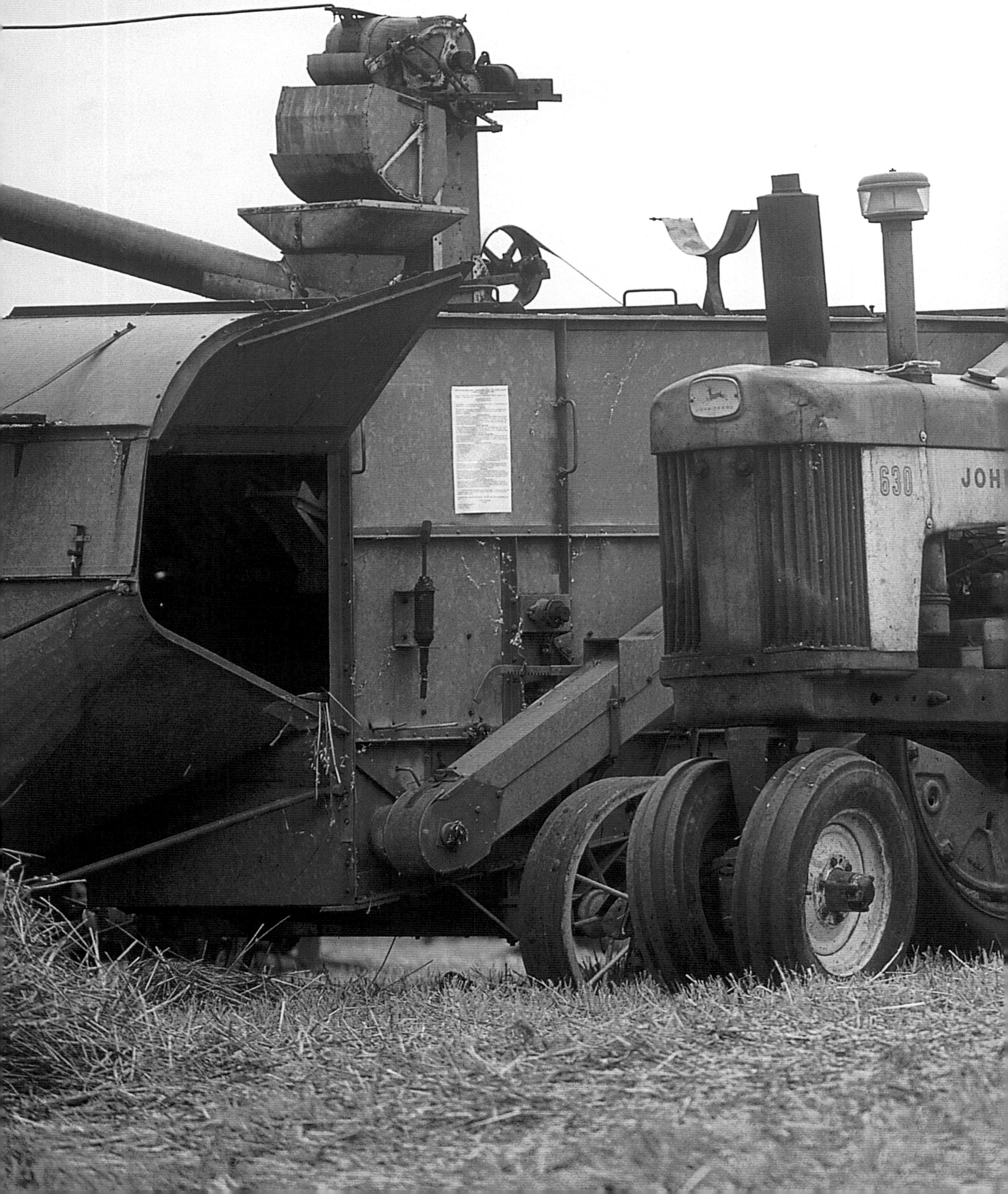
630
JOHN

A row crop John Deere 630. The 30 series was introduced in 1958 and ran on from then. The rear hubs are infinitely adjustable to suit row crop work.

Top: A 1963 52hp Fordson Super Major fitted with a Lupat mid-mounted mower.

Above: The mower is operated by a system of hydraulics both to position the mower and power the cutter. The mower is controlled by a number of levers positioned adjacent to the driver's left hand.

Right: This John Deere has an additional weight below the radiator grille to assist with traction.

Next page: One of the numerous 10 series tractors, this is a high clearance version of the John Deere 4010 powered by a diesel engine.

JOHN DEERE
1962 "4010"
High Crop
Owned By
Robert Manley
MYAKKA CITY, FLORIDA

#142

JOHN DE

Previous page: A Ford 4400 fitted with a hydraulic front loader and less aggressive tyres.

Left and Below: The gradual evolution of Ford tractors is illustrated by these two vehicles. This (**left**) 1950 Industrial Fordson was once owned by Leeds City Council; the later Ford 4000 (**below**) is still in use on a dairy farm.

Right: John Deere diversified into smaller tractors during the early 1960s. The ride-on, lawnmower concept, as shown here, has now become widely accepted and many manufacturers produce such machines.

Below right: An orange Allis-Chalmers D272 and Perkins-powered grey Ferguson showing off their vastly contrasting colour schemes.

A hard worked and slightly modified David Brown tractor on the Mediterranean Island of Malta.

North America

70s

During the 1970s **White** produced a tractor named the Plainsman. This was an eight-wheeler powered by a 504cu in displacement engine that produced 169hp.

Noted makers of other large tractors, the Steiger company had been incorporated by the late 1960s and had production facilities in Fargo, North Dakota. By the 1970s Steiger was manufacturing the distinctive lime-green painted tractors. They were constructed around engines that produced up to 525hp. The tractors were named after animals including the Puma, Bearcat, Cougar, Panther, Lion and Tiger and each denoted a specific horsepower class.

By this time the **International Harvester** Farmall 966 was a 100hp machine with a transmission of 16 forward and two reverse gears and a 414cu in (6,784cc) displacement engine.

The renowned English tractor maker, David Brown Ltd was taken over by **Case** in 1972. This acquisition was seen as important by the American corporation as the English company had a large distribution network in Great Britain, outlets in Europe and even a number in the United States. From then on David Brown tractors would be painted in David Brown white and Case red.

In 1974 new agricultural products from Case including the 2670 Traction King tractor, the 6000

Far left: Restored Fordson at an English vintage rally.

Left: French four-wheel drive tractor ploughing in northern France.

Above: T. Gerrand's restored 1970 66hp International 634 at the Masham, England, vintage rally.

Right: Another International 634—this one a restored vehicle dating from 1969 seen on a Yorkshire farm.

Left: Allis-Chalmers in a tractor-pulling event.

These pages: 1960s tractors from three nations; an American International Harvester (**above**), a Romanian Universal (**right**) and an English Muir Hill 101 (**below right**).

Next page: A diesel powered David Brown 990 Implematic tractor.

series of moldboard ploughs and the F21 series of wheeled tandem disc harrows. A year later the 1410 and 1412 tractors appeared. There were further new tractors a year later, when the 1570 AgriKing and 2870 Traction King 300hp four-wheel drive machines were announced. As it was America's bicentennial year, Case also offered a limited edition version of the 1570 model known as the 'Spirit of '76'. The 70 and 90 Series tractors were new for 1978.

In 1976 **Massey-Ferguson** introduced the 1505 and 1805 machines. These both featured a 174hp Caterpillar V8 diesel engine. From 1970 onwards there were broad expansions throughout the organisation with major advancements in all product lines, worldwide market development and a programme of capital investment to enlarge and improve facilities. More than $1.5 billion was

invested in this program between 1975 and 1981. In 1978 sales had reached $4 billion which meant a quadrupling of sales in the previous ten years.

During the 1970s the **John Deere** 40 Series tractors were popular as by now they incorporated V6 turbocharged diesel engines that produced in excess of 100bhp. This latter figure was exceeded by the end of the next decade when John Deere's 4955 Model tractor produced 200bhp.

Versatile Manufacturing Ltd. changed ownership in 1976 and changed its name in 1977 to Versatile Farm Equipment Company, a division of Versatile Corporation.

Left: A John Deere 7520. This large tractor, seen here with a seed planter, was developed during the 1960s to maximise productivity on large farms.

Right: A hard worked Massey Ferguson on the island of Malta.

Below: An Oliver built by White who retained the Oliver brandname after its 1960 takeover.

Europe

During this decade things would change considerably for the innovative **English** Roadless company. The first major change was brought about by the death of Lt-Col Philip Johnson in November 1965 at the age of 88. Johnson had been active within Roadless right into his latter years and had only relinquished the Managing Director's post in 1962. His death meant that a reshuffle took place. Four-wheel drive tractor production continued with Roadless conversions being supplied for Ford's new range of 2000, 3000, 4000 and 5000 tractors. The 5000 was the most powerful, a 65hp unit with a choice of eight- or ten-speed transmissions and in Roadless converted form known as the Ploughmaster 65. This was followed by the Ploughmaster 95. These tractors, like their predecessors, had smaller front wheels than rear ones

These pages: 1960s and 1970s tractors vary in size considerably—ranging from the small garden type (**above**), to the large David Brown 1210 (seen **right** at a UK preservation event in the 1990s) and the International 656 (**below**).

but the company also developed four-wheel drive tractor conversions that had equally sized wheels all round. The Roadless 115 was one of the first of this design of tractor to be produced. Successful demonstrations of Roadless' products at the Long Sutton, Lincolnshire tractor trials meant that through the 1970s the company was not short of work. Increasingly stringent legislation affected the Roadless tractors as it did other makers; from the end of the 1970s its machines had to be fitted with the 'Quiet Cab'.

In 1979 the company relocated from its original premises in Hounslow to Sawbridgeworth near Harlow in Hertfordshire, England. Sales of Roadless 4x4 tractors had started to slacken, simply because of increasing competition from other 4x4 tractor makers including cheaper machines from Italian Same, and Belarus and Zetor from Eastern Europe as well as 4x4 machines from

Above: As part of a restructuring, Tenneco Inc merged with Case and International Harvester. This IH tractor was manufactured before the merger.

Left: An IH tractor equipped with a hydraulic front-loading shovel.

Below left: The specific requirements of the handling and loading of farm produce has led to the development of specialised machines.

Below: An Oliver 1955 with provision for extra weight over the front axle.

the other major tractor makers. Roadless diversified into log-handling machinery and forestry tractors. A combination of circumstances and the recession of the early 1980s forced the companies that made up Roadless into voluntary liquidation.

In **Italy** in 1973 Landini started production of production of the renowned 6500, 7500 and 8500 tractors With this launch of the 500 series of two- and four-wheel drive machines in this year, Landini advanced its wheeled tractors by applying engineering that was to become standard for years to come. Once the four-wheel drive models were proven, work began on the design of high-powered machines, aimed at producing power in excess of 100hp using in-line six-cylinder Perkins engines. The range was widened to give tractors with power outputs from 45hp up to 145hp. In 1977 production of this new series of tractors commenced.

Ferrucio Lamborghini sold all his companies when he retired; in 1972 the car and tractor portions of Lamborghini's business had been split, the tractor production operations were taken over by Same which continued to develop the Lamborghini range and increased the volume of production.

In 1974 Fiat Macchine Movimento Terra entered into a joint venture with the American tractor manufacturer Allis-Chalmers Corporation, which was called Fiat-Allis. In Italy Fiat Trattori S.p.A. was also founded and 1975 Fiat Trattori became a shareholder of Laverda. In 1977 Fiat Trattori acquired Hesston, as a way of gaining entry into the North American market. Fiat Trattori later took over Agrifull, which specialised in the production of small and medium size tractors then in 1984 Fiat Trattori became Fiatagri, Fiat Group's holding company for the agricultural machinery sector.

Left: An Oliver tractor fitted with a John Deere hydraulic front loader.

A change of ownership of the **German** Hanomag company in 1971 caused the cessation of tractor manufacture. The Claas company started the development of its range of sugar cane harvesters in the 1970s and in 1975 Braud produced its first grape harvester and went on to specialise in this field.

Valmet, a **Finnish** tractor maker, amalgamated with Volvo BM from **Sweden** in 1972. Since then the company has specialised in forestry tractors such as the Jehu 1122. This machine is a wheeled tractor although provision has been made to install tracks over the tyres to improve traction in deep snow and difficult going. Volvo BM Valmet also made the 2105, a 163hp six-cylinder, turbo-diesel tractor as part of its range.

Above left: A Fendt Favorit ploughing in Germany where it was made.

Left: Two variations of the compact tractor; a Ford for ground maintenance and a mini-excavator with hydraulic front shovel and back hoe.

Above: A 1973 Massey-Ferguson 135 during a break in haymaking.

Top: Much modified copmpact in farm condition.

Previous page: Farmhand equipment such as this hydraulic front loader is now made by part of AGCO, the Allis Gleaner Corporation.

Top: An early 1970s Massey-Ferguson 135 with factory cab.

Above: A Dutra tractor made in Hungary by the state-run company.

Right: A Ford 4000 showing signs of corrosion from the salt water and spray it encounters in its working life by the sea, where it is used to haul fishing boats up the beach.

Previous page: A 1960s IH McCormick at a ploughing match in the English Cotswolds.

Above: During the 1970s the styling of tractor cabs became more angular.

Top: The Series III Land Rover, introduced in the early 1970s, was aimed at farmers as both a farm runabout—as seen in the foreground fitted with spraying equipment—and as an agricultural machine, as exemplified in the background.

This page: The Porsche Junior tractors were made in Germany; a few were exported to England in the early 1960s. This Junior is powered by a single-cylinder 12.5hp diesel engine and now part of the Coldridge Collection.

These pages: During the 1950s Porsche took over Allgaier and made a range of tractors with common parts until 1964. The larger tractors had fluid flywheel clutches, draught control, multiple PTOs and differential locks. This restored Porsche is now part of the Coldridge Collection.

These pages: A Lister Goldstar Mark II tractor from 1961. It is powered by a 24hp Lister HB2 engine and ZF five-speed gearbox with differential lock, and is one of four prototypes. It is now part of the Coldridge Collection.

These pages: A Massey-Ferguson 35X that is now part of the Coldridge Collection. This MF-35X is a vineyard tractor and was manufactured in 1962. It is powered by a 41.5hp diesel engine and was used by a market gardener in Taunton, Devon, England from new, until it was acquired by the collection for restoration. It is fitted with a post hole borer and seen here with Mike Thorne, proprietor of the collection, at the wheel.

Overleaf and these pages:
A 1967 Massey-Ferguson 175 diesel
that is now part of the Coldridge
Collection. The MF-175 was, for a
time, the most powerful tractor in
the Massey-Ferguson 100 range.
Powered by a Perkins A4 236
engine it produced 66.4-bhp. This
particular tractor's original owner
was Berkshire Agricultural College.

These pages: A 1967 BMC Mini diesel tractor. This is a 4200, produced after a six-year period of development. Several Ferguson engineers were involved in the project and a number of Ferguson patents incorporated in the design. As a result there are a number of similarities with the Ferguson TE20. The Mini made 15hp at the PTO but never sold in large numbers. This particular BMC Mini is now part of the Coldridge Collection.

These pages: A 1971 Massey Ferguson 165 with a 4x4 conversion by Four Wheel Traction Ltd. This company offered two 4x4 conversions for the MF-165; one used the same sized wheels all round through the use of lift-up hub reduction boxes, while the other (seen here) used different diameter front and rear wheels. This MF-165 is now part of the Coldridge Collection.

This page: Now part of the Coldridge Collection, this 1976 MF-20 is powered by a 45.5hp diesel engine. It is the industrial version of the MF-135 and was once the property of the British Army. It features dual braking circuits, stop lights, horn, indicators and a speedometer..

6 THE 1980s

The farm crisis of the early 1980s was to have far reaching consequences within the American agricultural industry and accelerated the succession of mergers and takeovers that characterised tractor and farm machinery manufacture in the latter part of the 20th century.

North America

International Harvester was an early casualty of the recession. In the 1980 financial year the company's losses exceeded $3,980 million, and the year after they were similarly high. In 1982 the situation worsened as losses totalled more than $1.6 billion. Despite this catastrophe, the losses were reduced to $485 million for 1983, although International Harvester could

Left: The farm crisis of the 1980s had far-reaching effects on the world's tractor makers.

not survive. In 1984 Tenneco Inc., who already controlled Case bought International Harvester's agricultural products division and formed **Case-IH**. Tractors built since this date have incorporated ideas from both companies. In the restructure at Tenneco, Case-IH became its largest division. The Case-IH range for 1986 included the Model 685L with a four-cylinder engine that developed 69hp, an eight-forward and two-reverse speed 4x4 transmission and Hydrostatic steering. A larger model in the same range for 1986 was the Model 1594, with an in-line six-cylinder engine that produced 96hp and drove through a four-range semi-automatic Hydra-Shift transmission.

In 1986, **Steiger** was acquired by Case-IH and to consolidate the merger, both companies' 4WD tractor lines were unified in both colour and name. The name Steiger was associated with high power tractors and in a survey of farmers in 4WD tractor markets, this name was cited as the most popular and well-known 4WD tractor brand. This led to the brand name later being revived for the Case-IH 9300 series of giant 4WD tractors.

The final **Allis-Chalmers** tractors were produced during the mid-1980s. In 1985 the recession-hit company was taken over by a West German concern. One of the last Allis-Chalmers tractors was the Model 4W-305 of 1985. It had a twin turbo engine that produced power in the

Below: By the 1980s Ford was still using the distinctive blue and white paint scheme that had become well known a decade earlier.

Above: A Ford 7710 in an English farmyard. Tractor cabs greatly enhance operator comfort and productivity.

Next page: A four-wheel drive Ford TW-15. Note the additional weights forward of the headlights to enhance traction in heavy field conditions.

region of 305hp, and a transmission that included 20 forward gears and four reverse. The purchaser was Klockner-Humboldt-Deutz AG, who renamed the tractor division **Deutz-Allis**. Production of tractors at West Allis, Wisconsin stopped in December 1985. Ownership by Klockner-Humboldt-Deutz was short-lived, and in 1990 the company was acquired by an American holding company—Allis Gleaner Co. (AGCO), who would soon rename the tractor producer **AGCO Allis**.

The world-wide recession also took its toll on **Caterpillar**, costing the company the equivalent of US $1 million a day and forcing it to reduce employment dramatically. During the later years of the decade the company's product line was considerably diversified. This diversification led the company back towards the manufacture of agricultural products. In 1987 rubber-tracked, crawler machines known as Challengers appeared in fields

and were considered to offer a viable alternative to wheeled tractors.

In 1986 **Ford** bought the New Holland Co. and a year later bought the Versatile Tractor Co. of Canada. Also in 1986 **White** joined with the New Idea Farm Equipment Co. and was itself subsequently acquired when the Allied Products Corporation bought in.

Earlier, in 1981, a Dallas, Texas-based corporation, the TIC Investment Corporation, bought **White** Farm equipment and continued with business under the acronym of **WFE**. The Allied Products Corporation of Chicago bought parts of WFE including the Charles City, Iowa tractor plant

Above: Four-wheel drive John Deere with bale grapple.

Left: A John Deere with an hydraulically operated tipping trailer in a Scottish field.

and stock. In 1987 Allied joined White with the New Idea Farm Equipment Co. into a new division known as **White-New Idea**.

Expansion of the **John Deere** credit business continued in 1984 with the acquisition of Farm Plan, which offered credit for agricultural purchases, the company's first financing of non-John Deere products. The John Deere Model 2040 of 1986 was powered by an engine of 238cu in (3,900cc) displacement, using diesel fuel and producing 70hp. It was one of the Series 40 of three tractors, the 1640, 2040 and 2140 models. Also made in the same year was the John Deere Model 4450 of 7,600cc displacement that produced 160hp and featured a 15-speed transmission with a part-time 4x4 facility. Diversification continued in 1987 when John Deere entered the golf and turf equipment market and then in 1988 the company diversified further into recreational vehicle and marine markets. In the same year a world-wide Parts Division was established to increase sales of spare parts to owners of both John Deere and

Above: Tractors are still vital to haymaking as shown in this photograph of a John Deere tractor with a hay conditioner.

other makes of equipment. In 1989 John Deere Maximizer combine harvesters were introduced.

East European **Zetor** tractors have been sold in the USA since 1982, and in 1984 American Jawa Ltd. took over the distribution. Zetor tractors are distributed in America through two major service and distribution centres, one based in Harrisburg, Pennsylvania and the other in LaPorte, Texas.

In 1981 **Ford** and majority stakeholder Nacional Financiera s.n.c. formed Fabrica de Tractores Agricolas S.A. (FTA) in Mexico.

Europe

In 1981 the tractor-producing part of the British Leyland conglomerate was sold to the Nickerson organisation to join **Marshall**. During the mid-1980s, as has been mentioned earlier, the well known **Allis-Chalmers** tractor brand became Deutz-Allis under the ownership of Klockner-Humboldt-Deutz, based in the German city of Cologne. Later this operation became part of the AGCO Corporation.

In the mid-1980s the 3000 series of **Massey-Ferguson** machines was made available with a turbo diesel 190hp six-cylinder engine. The MF-398 was a smaller Massey-Ferguson tractor, powered by a 236cu in (3,867cc) diesel engine and featuring a 4x4 transmission. Two other Massey-Ferguson tractors from 1986 were the Models 2685 and MF-699. The former has a 353.8cu in (5,800cc) Perkins Turbo diesel engine that produces 142hp and is one of the 2005 series of three tractors, 2645, 2685 and 2725. Each has a four-wheel drive transmission that incorporates 16 forward and 12 reverse gears. The MF-699 was the most powerful in the MF-600 series of four tractors, MF-675, 690, 698T and 699. It has a 100hp engine.

The 1980s saw **Landini** both diversifying and specialising its products, and co-operating with Massey-Ferguson. Having identified the requirements of the growing wine and fruit production sector, Landini started production of a series of Vineyard, Orchard, Standard and Wide tractors, the V, F and L machines. Landini soon had a 25 percent share of this worldwide market and was established as a leading manufacturer of these products. In 1986 Landini launched a new Vineyard tractor range which allowed the Fabbrico Factory to become the sole supplier of Orchard and Vineyard versions of both crawler and wheeled tractors branded as Massey-Fergusons. In 1988 the company launched a redesigned series of tractors, the 60, 70, 80 series. This was followed by the com-

pany achieving a sales record with more than 13,000 tractors retailed. In 1989 Massey-Ferguson sold 66 percent of its Landini shares to the Eurobelge/Unione Manifatture Holding Co.

Also in Italy, **Same** tractor production continued and included the Condor 55 and Models 90 and 100 of the mid-1980s. The Condor 55 was a conventional tractor powered by a 172.4cu in (2,827cc) 1003P direct-injection air diesel, with three cylinders, producing 55hp. The Models 90 and 100 are from a range where the numerical designation approximates to the number of horsepower that they produce. Each has 12 forward and three reverse gears and can achieve 19mph. Both have two-speed PTOs and hydrostatic steering.

In 1983 Italy's other major tractor producer, **Fiat Trattori**, entered a joint venture with the

Below: This Massey-Ferguson is seen during the haymaking in Snowdonia, Wales during the harvest. The trailer is used to move bales.

Pakistan Tractor Corporation in Karachi, Pakistan; it was named as Al Ghazi Tractors Ltd. In 1984 Fiatagri acquired 75% of Braud shares. Within this major restructuring, Hesston and Braud joined forces in a new company, **Hesston-Braud**, based in Coex, France; it later became part of New Holland Geotech. Things changed again in 1988 when all of Fiat-Allis and Fiatagri's activities were merged to form a new company, **FiatGeotech**, the Fiat group's farm and earthmoving machinery division.

UK production of **Ford** tractors was carried out at Basildon in Essex, one of Ford's eight manufacturing plants around the world. The company exported to 75 countries around the globe. In 1986 the Ford acquired Sperry New Holland and merged it with Ford Tractor Operations and named the new company **Ford-New Holland**. In the same year in England Ford was producing tractors such as the Model 4610, with an engine of 201cu in (3,300cc) displacement that produced 64hp. It was one of 11 tractors in the Series 10,

which consisted of 11 tractors ranging from the 2,900cc 44hp Model 2910 to the 402cu in (6,600cc) 115hp Model 8210. Also in series was the 268.4cu in (4,400cc) displacement, 103hp Model 7610.

Frazier is a small specialist company based outside York in Yorkshire, England and typical of many companies that produce specialised farming machines. During the 1980s the IID Agribuggy was assembled from a number of proprietary components including axles, suspension and engines; it was intended as the basis for crop-spraying and fertiliser-spreading tasks. The machines were purpose-built and offered in both diesel and petrol forms. The diesel variant was powered by a four-

Previous page: A four-wheel drive Massey-Ferguson 590 with a tipping trailer.

Below: For many years New Holland specialised in combine harvesters but various mergers now mean that it is a major tractor producer.

Right: Massey-Ferguson was formed through the merger of Massey-Harris and Ferguson and later became part of the AGCO conglomerate.

cylinder Ford engine of 98cu in (11,608cc) displacement. The Agribuggy is designed to be adaptable and available in both low ground pressure and row crop variants.

While **Land Rover** products were still made and widely exported, the emphasis of the range was now on sport utility vehicles. The only exceptions were the Defender Models 90 and 110, refined versions of the original Land Rovers and still popular for agricultural applications. There have been numerous specialised agricultural conversions made to Land Rover vehicles during the course of production to make them suitable for specific farming tasks, such as carrying and operating implements.

The German made **Mercedes** Unimog has been developed into the Unimog System of vehicle and implements, based on the wide range of vehicles and an almost unlimited number of implement attachment options. Several ranges are offered from compact to heavy duty models and each has at least three standard implement attachment points and a hydraulic system that ensures implement operation. The Unimog has been designed so that its wheels distribute pressure evenly on the ground in order to minimise soil compaction. It is of all-wheel drive configuration to ensure optimum traction.

An equally advanced Mercedes-Benz tractor was the MB-trac 1500 introduced in 1986. This tractor featured a turbo diesel engine of 346cu in (5,675cc) displacement that produced power in the region of 150hp. The MB-trac was designed as a dual-directional tractor.

Previous page: For a period Fiatagri produced tractors including those badged as New Holland, such as the four-wheel drive model see here.

Below: David Brown was acquired by Case and tractor production continued, albeit with a change of colour scheme.

Right: This four-wheel drive Fiatagri 88-94 tractor also has a New Holland badge on its radiator grille.

Steyr, based in St Valentin, Austria, through the 1980s and 1990s offered a range of row-crop and utility tractors raging from 42 to 145hp.

In Eastern Europe, based in the former USSR, **Kirov** produced tractors, including the giant K-701s. These were 12-ton machines powered by a liquid-cooled V12 diesel producing 300hp and driving through a semi-rigid coupling and a reduction unit which can be disconnected to facilitate engine starting in cold weather. The frame of the K-701 is articulated to aid traction.

The **Zetor** 8045 Crystal Model of 1986 had an engine of 4562cc displacement and used diesel fuel. It produced power in the region of 85hp. It is a 4x4 tractor with eight forward gears and two reverse ones.

The **Belarus** Model 862D of 1986 was powered by an engine that displaced 4,075cc and produced 90hp. It featured an automatic, four-wheel drive transmission that incorporated 18 forward and two reverse gears. Hydrostatic steering was fitted as standard.

Above: The increasing size and power of tractors has meant that two field operations can be carried out at once by mounting implements on the front and rear of tractors.

Above right: Harvesting potatoes during October in mid-Wales.

Right: A four-wheel drive Mercedes tractor from the early 1980s—the MB-trac 800.

Previous page: An International 986 tractor.

Another East European company, **Universal** was a Romanian concern based in Brasov. It began production in 1946 and exported widely. A mid-1980s' product was the Model 1010 powered by a 100hp 5,393cc six-cylinder diesel engine. In the USA the Romanian Universal tractors were marketed under the Long brand name.

Rest of the World

The Japanese company **Kubota** was founded in the last decade of the nineteenth century. It began manufacturing tractors in the 1960s and claimed to be the fifth largest producer by the mid-1980s.

Above: The MB-trac 800 is powered by a 3,780cc four-stroke diesel engine that produces 75hp

Right: The Frazier Agribuggy is a specialist tractor made in England and suited to low ground pressure uses such as crop-spraying. The late-1980s version pictured is equipped for this task.

Previous page: The Mercedes MB-trac 1500 of the mid-1980s was dual-directional and powered by a 150hp engine.

Above left: The Highlander is a specialised four-wheel drive forestry tractor, note protection bars around the cab.

Below left: Machines such as this Matbro are used as loaders in farming situations.

Above: The Agrover was a specialised agricultural conversion to the standard Land Rover 110 to make it more suited to farming applications.

Below right: The Agrover was fitted with a live PTO at the rear.

Previous page: To load round hay bales onto trailers for transport from the field the farmer needs a bale grapple. This Ford tractor, seen in Wales, has both a bale grapple and a back hoe fitted.

Above and Right: The rear body of the Agrover was designed to be removable to allow other equipment to be installed. Ground clearance was enhanced through the fitting of chain-driven portal axle-type hubs.

Next Page, Main Photo: Kubota is a Japanese maker of compact tractors. This L3750 model is fitted with a rear mounted mower.

Next Page, Inset: The handling and storing of round hay bales requires specialised implements.

Right: The Japanese Kubota L2250 is a small tractor intended for lawn and ground maintenance.

Below: The German Mercedes-Benz Unimog is a multi-purpose machine popular for numerous agricultural applications.

7

THE 1990s ONWARDS

North America

In 1990 Klockner Humboldt Deutz Ag its Deutz-Allis division to **AGCO** (an acronym for Allis-Gleaner Company). Since then AGCO Corporation of Waycross, Georgia has become a multinational farm machinery company through a combination of market growth and acquisitions. In the same year the company began manufacturing and distributing farm equipment under the AGCO Allis and Gleaner brand names.

In 1991 AGCO purchased both the Hesston Corporation and the tractor division of White-New Idea, although the company retained the White name as a specific brand of tractors. In 1992 AGCO purchased the North American distribution rights to the Italian Same brand of specialist tractors and followed

Left: The Case-IH logo came about following the merger of Case and International Harvester.

this in 1993 by buying the distribution rights to Massey-Ferguson products from the Varity Corporation. AGCO acquired the remaining portion of White-New Idea during 1994 and purchased the world-wide holdings of Massey-Ferguson. In the same year AGCO purchased McConnell tractors, subsequently leading to the development of the AGCOStar articulated tractor line and Black Machine planter equipment. By 1995 AGCO had acquired the distribution rights to Landini tractors for North America and purchased the Ag Equipment Group, the makers of Glencoe, Tye and Farmhand equipment.

Acquisitions continued in 1996 when AGCO acquired the Iochpe-Maxion agricultural equipment company in Brazil and two Canadian companies—the Western Combine Corporation and Portage Manufacturing Inc. Two more firms were purchased during 1997, Deutz Argentina S.A. and Fendt. The latter was a major German tractor

Above and Right: Ford and New Holland worked closely together in association with Fiat. The Ford name has now disappeared from tractors.

company. The AGCO corporation has numerous production facilities around the world. In the USA these include Duluth, Georgia, Batavia, Illinois, Hesston, Kansas, Independence, Missouri, Coldwater, Ohio and Lockney, Texas. In Europe there are plants in Coventry, England, Beauvais and Ennery, France, Baumenheim, Kempten and Marktoberdorf, Germany. In South America production is carried out in Queretaro, Mexico, Canoas and Santa Rosa, Brazil, Haedo, Noetinger and San Luis, Argentina. The company also has a plant in Sunshine, Australia.

The range of AGCO Allis tractors for the 1990s included several series of different models including those designated as 5600, 6670, 8700 and 9700. The 5600 series models are mid-sized machines and range from the 45hp 5650 to the 63hp 5670. They are powered by direct-injection air-cooled diesel engines. The 5600 models are designed to be capable of tight turns and to drive a wide variety of equipment from the PTO. The given horsepower figures are measured at the PTO. The AGCO Allis 6670 tractor is a row-crop tractor that produces 63hp from the PTO (63 PTOhp). It is powered by a four-cylinder direct injection air-cooled diesel engine. Reactive hydrostatic power steering is designed to end drift and what is termed the 'economical PTO' setting is designed to save fuel by reducing engine speed by 25 percent on jobs that do not require full PTO power. Models 8745 and 8765 are large capacity tractors designed for large acreage farming. The 8765 is available with a choice of all-wheel drive or two-wheel drive and produces 85hp at the PTO, while the 8745

Right: A four-wheel drive Case tractor fitted with a hedge trimming attachment.

produces 70 PTOhp. Both come with 12-speed synchronised shuttle transmissions which are designed to offer sufficient power in all gears. These tractors have an AGCO Allis 400 series turbocharged diesel engine providing the power under the bodywork which is of a modern 'low-profile' appearance. The larger tractors in the 8700 series are the Models 8775 and 8785 powered by a fuel-efficient AGCO Allis 600 Series liquid-cooled diesel. These tractors produce 95hp and 110hp at the PTO respectively. They are manufactured with a choice of all-wheel drive and two wheel drive transmissions. The transmission has four forward speeds and an optional creeper gear. They are also equipped with a 540/1000 'economy PTO', a durable wet multi-disc clutch and electronically controlled three-point hitch. The 125 PTOhp 9735 and 145 PTOhp 9745 models are tractors that have been designed with styling and performance

Above and Right: The variety of farm tractor duties has not diminished over the years as can be seen by these photos: a Ford spraying a seeded field (**above**); another Ford in an English farmyard in front of a laden trailer (**right**); and a four-wheel drive New Holland fitted with a Quicke 430 front loader fitted with the attachment for lifting pallets.

in mind. Both the models are available in an all wheel drive configuration or as a standard model with either four or 18-speed transmissions. The two turbocharged 9600 Series 'Powershift' models are fitted with a liquid-cooled engine of up to 8,700cc (530cu in) displacement. The 175 PTOhp Model 9675 and the 195 PTOhp Model 9695 tractors feature an electronic transmission—called Powershift by its makers—and are of a cab forward design. Computer technology has been

employed in the manufacture of the AGCO-Allis range. The proprietary system—DataTouch—is designed to be a compact, easy-to-read, touch-screen display from which all of the functions of the in-cab systems can be controlled. It has been designed so that there are no cables in the operator's line of sight. The system is based on simple touch-screen technology that is considered to be easy to use despite the complexity of the operations and data it controls.

The Gleaner company, part of the AGCO Corporation and a manufacturer of combine harvesters for more than 75 years, currently offers a range of four rotary combine models as well as the C-62 Conventional Combine all of which are powered by Cummins liquid-cooled engines. Gleaner combine harvesters are built on a one-piece, welded mainframe and feature a centre-mounted

'Comfortech' cab. The combine harvesters are designated as follows, R42, R52, R62, R72—the R series rotary combines—and C62—the conventional combine harvester. The R42 Gleaner Class four combine, has a 170-bushel grain tank, a 100-gallon fuel tank and a 185hp Cummins engine. The R52 Gleaner Class five combine has a standard 225-bushel grain tank and 245-bushel option. Its polyurethane fuel tank has a 100-gallon capacity. A 225hp Cummins engine is fitted. The R62 Gleaner

Left: Tractors frequently have long working lives as evidenced by this ageing Massey-Ferguson in an English farmyard and still in use.

Above: Tractor cabs offer protection from the cold as well as the heat. This Ford in a winter field has both a cab and a front loader.

Right: Tractor cabs are becoming increasingly ergonomically designed in order to increase visibility and ease of operation.

Above: The design of tractors is becoming increasingly curved as shown by these contemporary Massey-Fergusons .

Right: A four-wheel drive Ford planting seeds with its rear-mounted planter.

Class six combine offers a 260hp in-line six-cylinder engine, and has as standard a 225-bushel bin or as an option a 300-bushel grain tank. The R62 offers accelerator rolls, chaff spreader, two distribution augers and a 150-gallon fuel tank. The R72 Gleaner is the only Class seven combine currently built in North America. It features a 330-bushel grain tank and a 300hp in-line six-cylinder engine. The R72 includes two distribution augers, accelerator rolls, a chaff spreader and a 150-gallon fuel tank. The C-62 Conventional Combine harvester has a 300-bushel grain bin and a high-capacity, turret unloading system that delivers grain into trucks and trailers at a rate of 2.2 bushels a second. The C62 is constructed around an all-welded frame using heavy-duty final drives and designed with

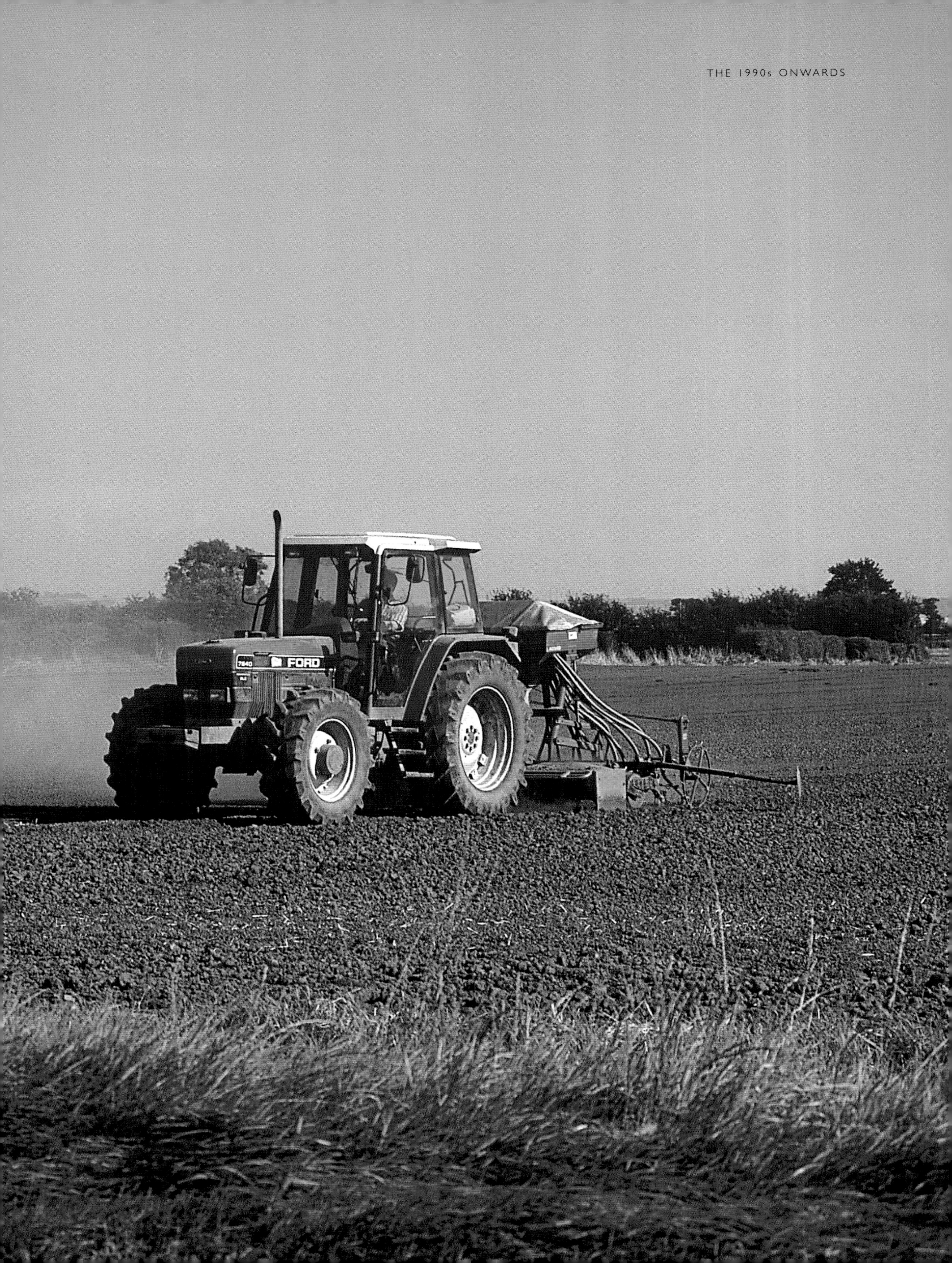
FORD
7840

Left: The progression in tractor design is illustrated by two generations of four wheel drive Massey-Ferguson; the newer Model 4263 in the foreground is noticeably more curved than the earlier model behind.

Above; A four-wheel drive Massey-Ferguson with a five-furrow plough.

even weight distribution and strength needed for large loads. Powering it is a 260hp Cummins diesel engine.

In 1993 AGCO purchased the **White-New Idea** range of implements although it retained the White name for their brand of tractors. The current White tractor models range from 45 to 215hp include proprietary systems such as Synchro-Reverser, Powershift and Quadrashift transmissions and are powered by Cummins direct-injected diesels. The range includes the following models, 6045 and 6065 Mid-Size, 6090 Hi-Clearance, six Fieldmaster and three Powershift models. The 45 PTOhp 6045 Mid-Size has a 50-degree turning angle and there is a choice of either two-wheel drive or a model with a driven front axle. The

Synchro-Reverser transmission has 12 gears in forward and reverse. The 6065 Mid-Size develops 63hp at the PTO and has a 12-speed Synchro-Reverser transmission and a four-cylinder engine. Both two- and four-wheel drive models are available. The 6090 Hi-Clearance tractor is an 80 PTOhp tractor with generous ground clearance. It gives 27.2in of crop clearance and 23.2in of ground clearance under the front and rear axles as well as 21.7in of ground clearance under the drawbar. The 6090 in this form has a synchromesh transmission with 20 forward and 20 reverse gears as standard. The 6175 and 6195 Powershift Series ranges from 124 to 200 PTOhp and offers an electronic-controlled full Powershift transmission. It has 18 forward and nine reverse gears. The White 6215 Powershift has 215 PTOhp and a Powershift transmission with 18 forward and nine reverse gears. It

is powered by a Cummins six-cylinder, 8,300cc displacement turbocharged aftercooled engine.

Massey-Ferguson became part of AGCO in 1994. According to the manufacturer at the time, 'for 33 straight years more people have purchased MF tractors than any other brand'. Massey Ferguson's 1990s range was comprehensive and included 130–180 PTOhp tractors, 86–110 PTOhp tractors, 55–95 PTOhp tractors, 34–67 PTOhp tractors, 37–53 PTOhp tractors, 16–40 engine hp

Below: Sheep shelter from the wind and snow of an English winter in the lee of a Case tractor.

Right: A tractor equipped with a hydraulic front loader returning from the upland fields of West Yorkshire, England.

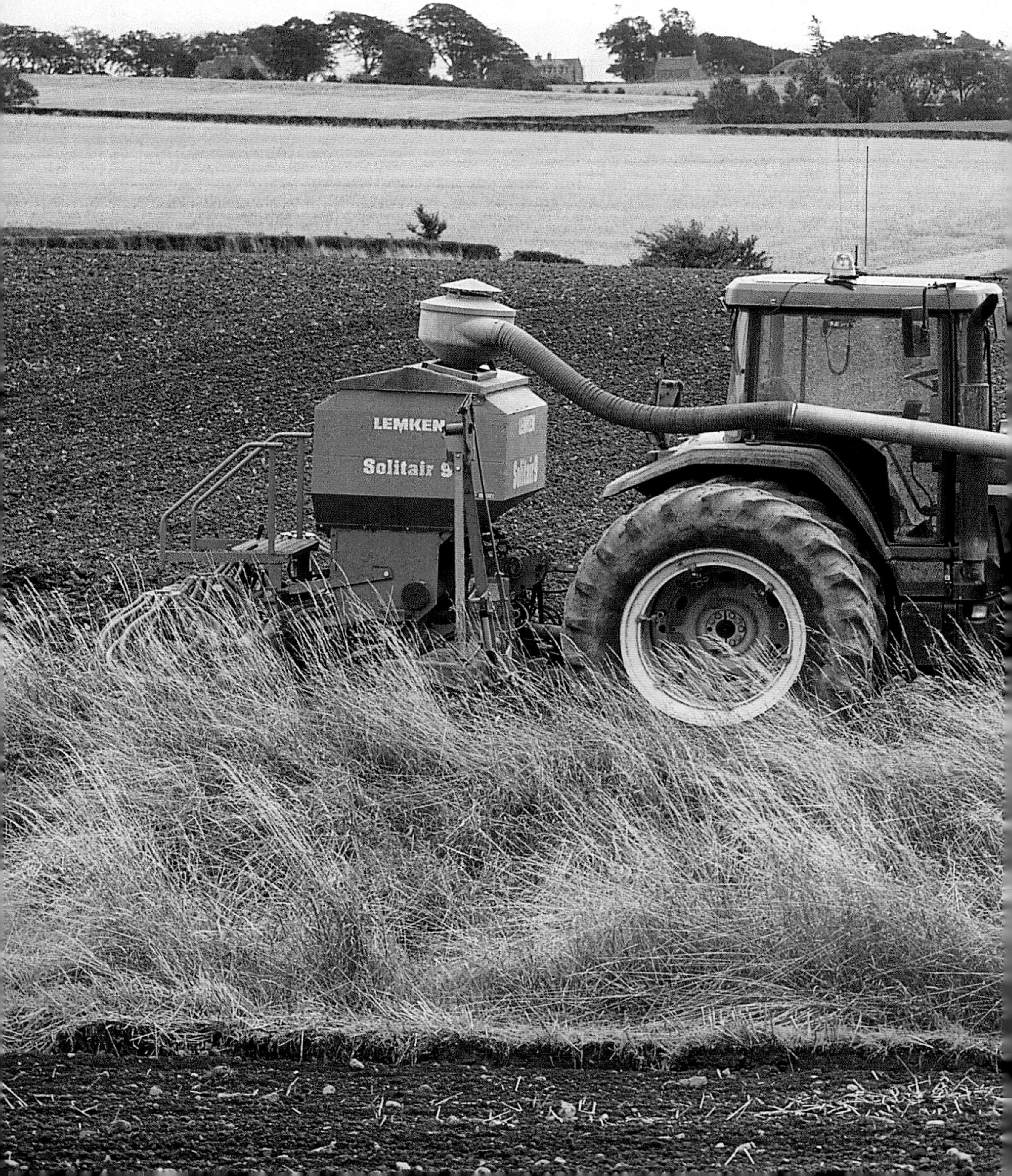
LEMKEN
Solitair 9

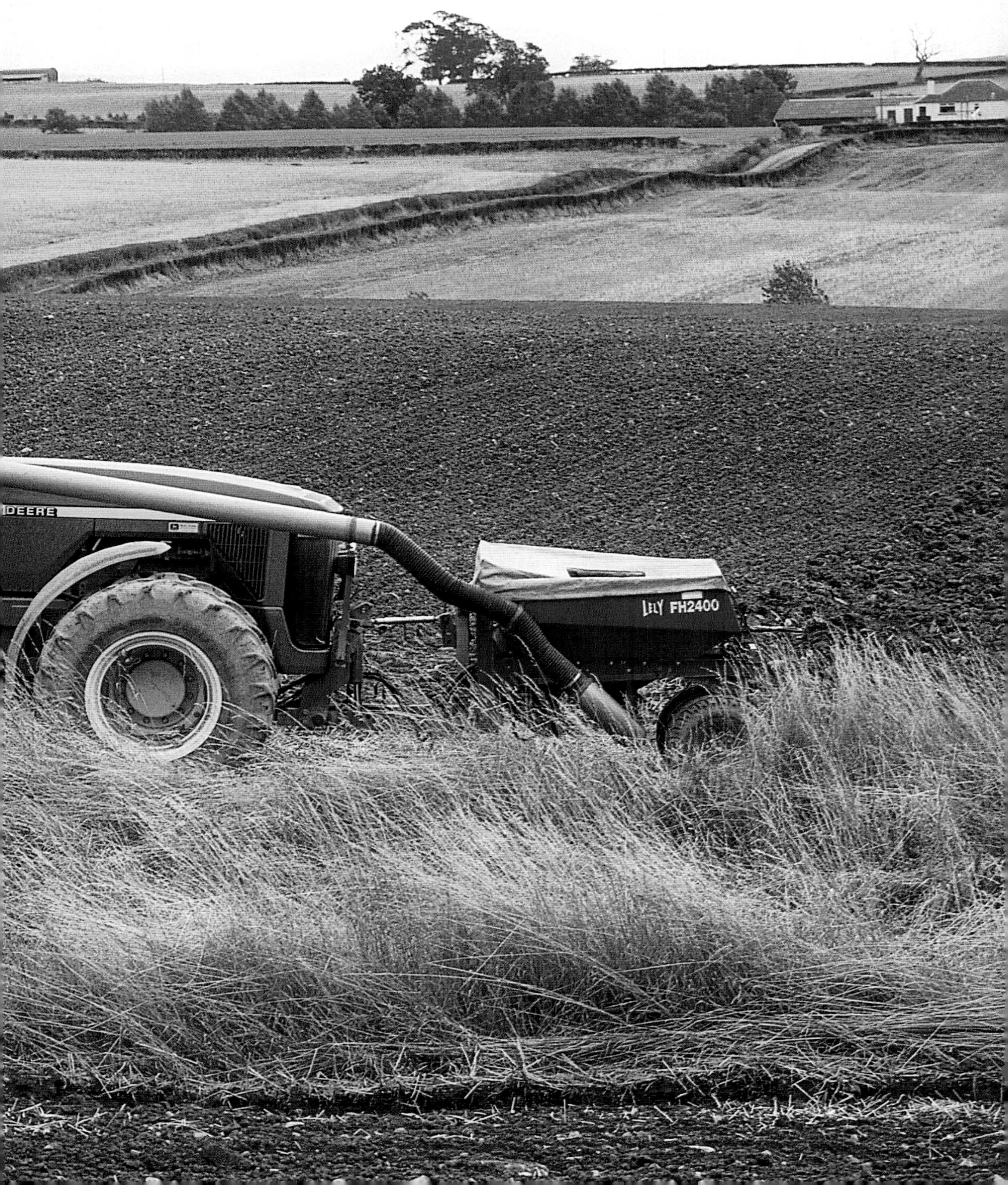
DEERE
LELY FH2400

Previous Page: A John Deere photographed during a break from planting in previously prepared fields.

tractors as well as conventional Combine harvesters, rotary combine harvesters and loaders. The 8100 series of 130–180 PTOhp tractors are designed as high-performance tractors while the 4200 55–95 PTOhp tractors are available as a selection of models of varying configurations, differing options and attachments. The 200 series of 34–67 PTOhp tractors feature Perkins diesel engines. Another small range of tractors from Massey-Ferguson was the Series 1200 16-40 PTOhp. Massey-Ferguson also manufactured Class five and six combine harvesters including the 8680 Conventional and 8780 Rotary Combines at this time.

AGCOStar—another and new, AGCO brand— tractors appeared in 1995 and among the first products were the Models 8350 and 8425. Both were state of the art mid-1990s machines that used the most up to date technology to provide a tractor suitable for large acreage farming. The AGCOStar Model 8425 of 1997 was powered by a 425hp Detroit Diesel engine linked to an 18 forward gears and two reverse transmission and centre articulating steering for increased manoeuvrability. The 360hp Model 8360 was similarly equipped and both were built on a C-channel high-strength front steel frame for durability. Later models of AGCOStar tractors featured the Cummins N14 Series in-line six-cylinder engines. These diesels are of 14 litres (855cu in) displacement and are both turbocharged and aftercooled and designed to combine a high torque rise with maximised fuel efficiency and reduced exhaust emissions. The tractors have constant mesh

synchronised transmission with 18 forward and two reverse gears. The gearbox provides eight speeds in the 3–8mph range and to enhance manoeuvrability the tractors articulate up to a 35-degree angle. Maintenance access is designed to be convenient and simple. A swing-out front grille allows the operator to reach the radiator, air conditioner evaporator and oil cooler. The air cleaner is positioned to the right of the cab. Dual or triple sets of tyres can be fitted when conditions require it to assure sufficient traction and flotation while minimising soil compaction. In order to enhance operator comfort the isolation-mounted ROPS cab is designed for comfort; the controls are located for maximum convenience while hydrostatic brakes, tilting and telescopic steering and a 'float-ride' seat are designed to make long days less tiring for the operator.

In 1991 **Fiat** acquired **Ford New Holland Inc.**, merged it with the already extant

Above Left: Low ground pressure tyres can be massive as illustrated by the ones fitted to the rear of this tractor.

Above: A John Deere fitted with a hydraulically operated front loading shovel.

FiatGeotech and renamed the new company **N. H. Geotech**.

In 1992 **Ford** Tractor in Brazil was moved from Sao Paulo to Curitiba. The Curitiba plant currently produces tractors as well as combine harvesters. In 1993 N.H. Geotech was renamed New Holland. It was specified that the Ford name can only be used by New Holland Inc. until 2001 by agreement with Fiat. At the corporation's second worldwide convention held in Orlando, Florida, New Holland launched 24 tractor models in three different ranges alongside the Fiat-Hitachi Compact Line. The Versatile 82 series range of articulated tractors are high-powered tractors designed for the biggest of fields and the heaviest applications. The

Left and Above: John Deere have stayed at the forefront of tractor innovation as evidenced by the curved glass incorporated into the cab of this 3350 model (**left**) and the increasingly curved bodywork of this four wheel drive model (**above**).

range included the Models 9282, 9482, 9682 and 9882 all of which are powered by six-cylinder engines of varying horsepower with twelve-speed transmissions in both two- and four-wheel drive configurations. The Series 70 tractors are also designed for large farms and have power outputs that range between 170 and 240hp. They incorporated New Holland's own 'Powershift' transmission which offered single lever control of the 18 forward and nine reverse gears. They were manufactured in Winnipeg, Canada and marketed worldwide. The tractors are available with cab, less cab or with ROPS (roll over protection system).

The third range of New Holland tractors was the 90 series models which have a range of transmissions. The 100-90 and 110-90 models were manufactured in Jesi, Italy while the 140-90, 160-90 and 180-90 models are manufactured in Curitiba, Brazil. In 1997 New Holland completed its purchase of Ford Motor Credit Co.'s partnership interests in the two joint ventures that provided financing for New Holland's products in the United States. New Holland also signed an agreement with Manitou for the design and production of a New Holland range of telescopic handlers.

In India, in 1998 New Holland completed the construction of a new plant for the manufacture of tractors in the 35–75hp range and commenced production. New Holland then signed an agreement with Flexi-Coil, a Canadian manufacturer of air seeding systems and tillage equipment.

In Turkey, New Holland made a new agreement with its partner, KoÁ Group, increasing its share in the joint venture Turk Traktors to 37.5 percent. New Holland Finance expanded its activities from the UK to other European markets including Italy, France and Germany.

During the 1990s and after a major restructuring Tenneco formed the Case Corporation with its agricultural products badged as Case-IH machines. The current Case-IH range includes Magnum and Maxxum tractors. The 8900 Magnum series tractors are powered by an 8,300cc (506cu in) displacement engine and assembled with an 18-speed Powershift transmission. There are four models in the MX Series of Maxxum tractors with outputs ranging from 85 to 115 PTOhp. They are suited to a variety of tasks including tillage, planting and loader work. They also designed to adapt to speciality, row-crop, hay and forage and utility applications. They offer the following PTOhp ratings: MX100, 85hp (63kW); MX135, 115hp (86kW).

The Maxxum series is fitted with a modern design of six-cylinder 5,883cc (359cu in) displace-

Right: Four-wheel drive tractors have become increasingly common and all the major makers offer at least one. This is a John Deere 3650.

Left and Right: Despite the improvements in tractor technology, including four-wheel drive transmission systems, the addition of weights forward of the front axle is still common to aid traction in field conditions. Both these four-wheel drive John Deeres are so equipped.

ment turbocharged diesel engine. The Maxxum tractor transmission is a 16-speed fully synchronised unit and is standard on all models except on the MX135 which has a Powershift transmission. Standard creeper and super creeper transmissions are available on all models. A compact series of three Maxxum tractors is also offered—the Models MX80C, 90C and 100C. Alongside these are also the CS110-150 and CX50 to CX100 models.

A particularly specialised agricultural machine manufactured by **Case-IH** at its East Moline plant is the 2555 Cotton Express, a state of the art cotton-picker. Cotton-picking is a specialised business and the machines designed to do it have become increasingly technical. The 2555 is capable of holding up to 8,500lb (3,864kg) of cotton and has a 1,150cu ft (32.5cu m) vertical lift basket. The design of components such as straight-through front and rear outlets are intended to enhance crop flow and so reduce chokes, improve drum component life and decrease wear on moistener pads and pad holders. The 2555 is powered by a 260hp (194kW) turbocharged diesel engine and offers on-road transport speeds of around 18mph (29kph) as well as sufficient power for harvesting. A more conventional, but no less technologically advanced, combine harvester currently made by Case-IH is the Axial-Flow 2100 series.

In North America Case-IH has numerous production facilities each dedicated to specific products. In Burlington, Iowa loaders, backhoes, forklifts, crawlers, dozers and hydraulic cylinders are produced. In East Moline, Illinois the company manufactures combine harvesters, cotton-pickers, grain and corn heads. In Fargo, North Dakota the Case plant produces tractors, wheeled loaders and Concord air drills. In the same state a plant in Valley City is dedicated to electronics. In Hamilton, Ontario, Canada tillage, crop production and material handling equipment are made. In Racine, Wisconsin Case produces its lines of Magnum and Maxxum Tractors as well as transmissions and foundry products. In Wichita, Kansas are made skid

steer and trenching machines; in Hugo, Minnesota directional drills are fabricated. In South America, a plant in Sorocaba, Brazil makes wheeled loaders, backhoes and excavators. Case also has a factory in Australia in Bundaberg, Queensland where Austoft sugar cane harvesters, transporters and planters are all made.

The corporation participates in a number of joint ventures including those with Hay & Forage Industries of Hesston, Kansas, producing hay and forage equipment, and the Consolidated Diesel Co. of Rocky Mount, North Carolina who make engines. Further afield, Case works with the Liuzhou Case Liugong Construction Equipment Co. Ltd. of Liuzhou, China to make loaders and backhoes. Case have a similar arrangement with Brastoft Maquinas e Sistemas Agroindustriais S/A in Piracicaba, Brazil to produce sugar cane harvesters, and with UzCaseMash in Tashkent, Uzbekistan where cotton-pickers are made.

Case's 1997 revenue amounted to a record $6 billion, up 11 percent over $5.4 billion in 1996. Net income reached a record $403 million. At this time the corporation had approximately 18,000 employees, and 4,900 dealers and distributors. Later in the decade Case-IH formed a Latin American agricultural equipment unit and announced it will invest up to $100 million over three years to manufacture large-scale, production agricultural equipment in Brazil for the Latin American market. Case-IH also acquired Agri-Logic, a supplier of yield-mapping software.

Left: A two-wheel drive John Deere Model 6210. Two-wheel drive tractors are usually fitted with ribbed tyres on their front wheels, as here.

Left and Right: John Deere offers a complete line of agricultural equipment including implements, as well as the two-wheel drive tractors (**left**), smaller machines intended for lawn and grounds maintenance (**below left**) and four-wheel drive tractors (**below**). Other manufacturers such as New Holland (**right**) also offer a comprehensive range.

Next Page: A New Holland 8560 with a sprayer. The spraying booms are seen here folded in to allow the tractor and implement to be driven to and from the field.

The 1990s range of **Steiger** tractors from Case-IH consisted of the 9300 series of massive four-wheel drive tractors. This series comprised a range of ten Steiger machines. The models ranged from 240 to 425hp, with two row-crop special models and the Quadtrac tractor which featured four independent crawler tracks. The current 9300 series tractors were designed and manufactured in Fargo, North Dakota. The 9300 series models range from the 240hp Model 9330 to the 425hp (317kW) Model 9390. The optional transmissions are as follows, a 12-speed SynchroShift standard, a 24-speed hi-low SynchroShift and a 12-speed full Powershift although not all these transmissions are available on all models. This series of tractors was comprehensively equipped in terms of hitches, hydraulics and PTOs. Lift capacities ranged from 12,750 to 17,000lb (5,783 to 7,711kg) depending on the model.

The 1990s were auspicious times for **John Deere**. The introduction of an entirely new line of 66–145hp tractors, the 6000 and 7000 series came

in 1992. These were designed from the ground up as a complete range of products to meet the varying demands of world-wide markets and to facilitate ongoing product updates quickly and at minimum cost. In 1993 the company's 75th year in the tractor business the tractor line spanned a range from 40hp to 400hp. In 1994 John Deere Specialty Managers, a new business unit within John Deere Insurance Co., was established to serve specialised markets for property and casualty insurance. The 8000-series tractors were claimed to set new standards for power, performance, manoeuvrability, visibility, control and comfort in 1994. The 8400 was the world's first 225hp row-crop tractor for example. In 1995 the company introduced a new line of mid-priced lawn tractors and walk-behind mowers branded as 'Sabre by John Deere' for distribution through John Deere dealers and national retailers. In that year it also announced the 'GreenStar Combine Yield-Mapping System' as the first in a series of precision-farming systems that was designed to help measure crop yield in

different parts of a particular field. The company made significant financial moves in 1995 when its stock was split three for one from November 17, 1995. The previous stock split was a two for one split in 1976. In this year the company's consolidated net sales and revenues exceeded $10 billion. Financial moves continued in 1996 when the Board authorised a $500 million share repurchase. For the year John Deere's consolidated net sales and revenues were a record $11.2 billion. In the same year the Lawn and Grounds Care Division name was changed to Commercial and Consumer Equipment Division.

While John Deere is primarily an American tractor manufacturer, the company has production plants in other countries including Argentina and Mexico. Other plants are further afield in Australia, South Africa and Europe—the latter in Germany and Spain.

The **Caterpillar** range for 1996 included a model called the Challenger 75C. Its engine displaced 629cu in and produced maximum power of 325hp. The operating weight is in excess of 16 tons. The rubber crawler tracks fitted are referred to as Mobil-Trac by the manufacturer. The company continued to expand and acquired the UK-based Perkins Engines in 1997. Considerable innovative technology is employed in the assembly of agricultural crawlers. The undercarriage is designed to transfer maximum engine power to the drawbar. Because there is less slip with tracks than wheels the crawler will do more work with less horsepower and require less fuel to do so. Traction under heavy loads is a variable; a wheel tractor

Below: The Caterpillar Challenger 55 uses Caterpillar's rubber crawler track technology known as Mobil-Trac.

Above: A John Deere 8850, with dual tyres to increase traction and reduce soil compaction, pulling a disc harrow.

Left: The John Deere 1560 No-till seed drill increases productivity by tilling and planting in one pass.

Below left: Combine harvesters such as this can be regarded as tractors specialised for one farming task.

Next page: Both Massey-Ferguson (**insets**) and John Deere offer state of the art tractors. While the Massey-Fergusons are seen ploughing the John Deere is pulling a planter.

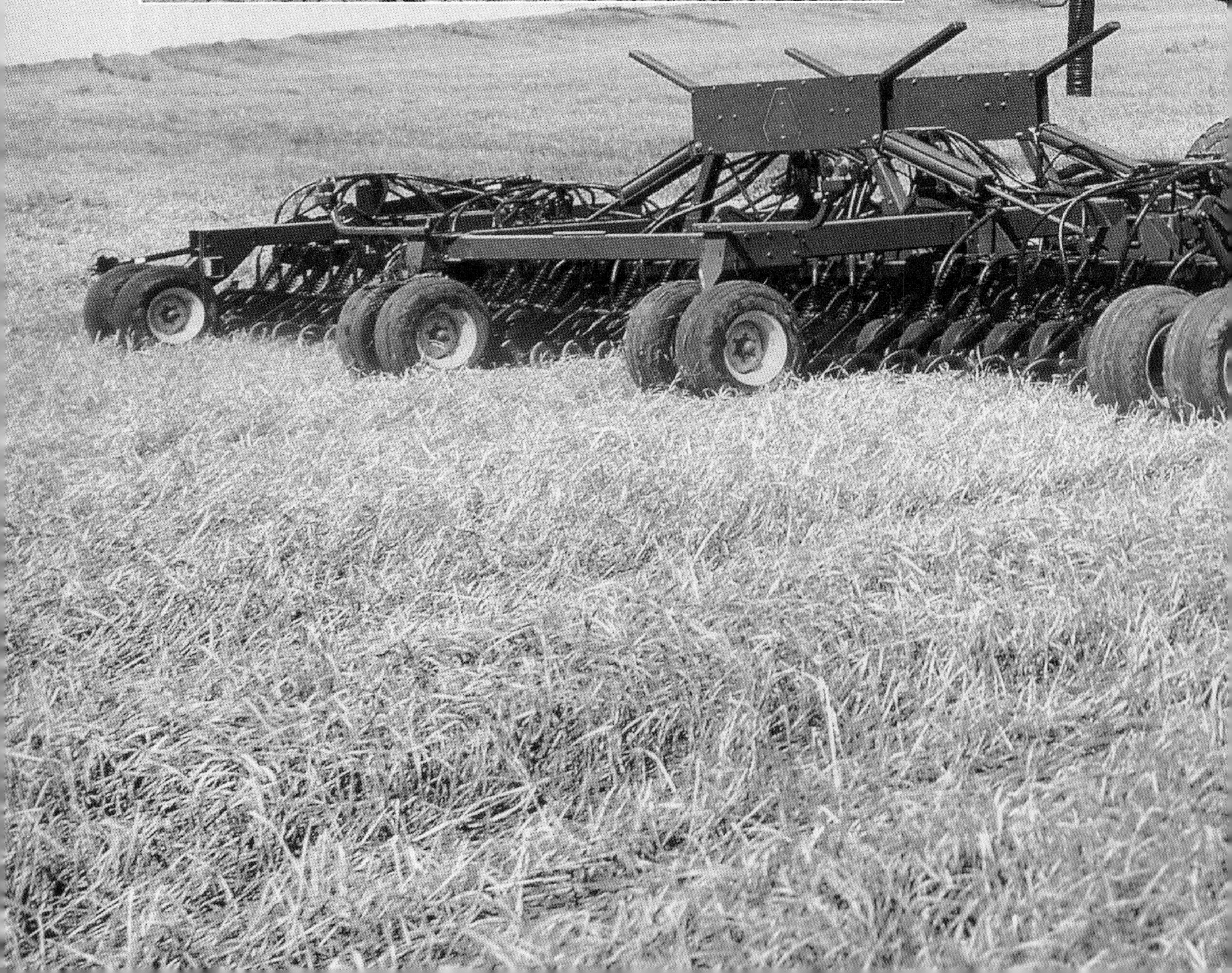

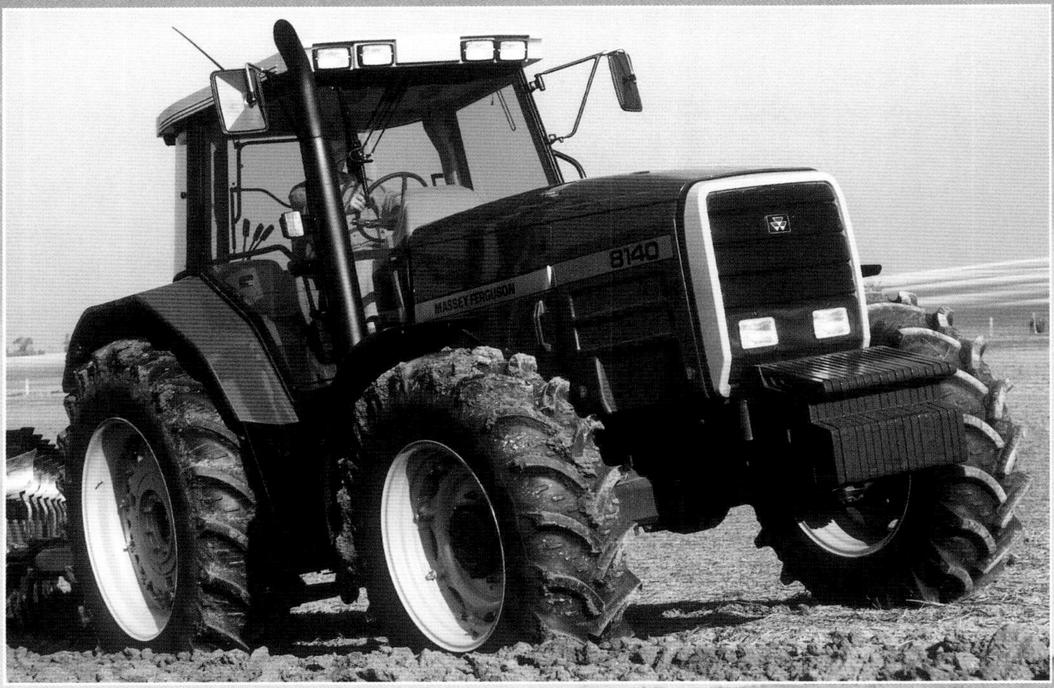
8140

Above: Massey-Ferguson offers a full range of farming equipment including towed balers such as this Model 130 square baler.

Right: Spreading fertiliser on a growing crop; tractors are designed to minimise damage in this situation.

typically experiences slip levels of up to 15 percent while under the same heavy drawbar loads, a Challenger tractor with Mobil-trac technology reportedly experiences only up to four percent of slip. The long, narrow footprint of the rubber crawler tracks allows the operator to get into the field in difficult conditions by both lowering ground pressure and increasing flotation. By distributing the gross weight over more axles, Cat's crawler tractors and combines achieve reduced soil compaction.

In late 1999 the Case Corporation and New Holland completed a merger. The company, which is traded on the New York Stock Exchange, is one of the largest in the equipment industry with combined 1998 revenue of approximately $12 billion. The multiple brands and corresponding distribution networks of both the Case and New Holland organisations will be maintained in the market-place. The merger agreement was initially announced by the two companies in May 1999. In late autumn, the merger received clearance from regulatory agencies in Europe and the United States. Case had 1998 revenues of $6.1 billion and at the time of the merger was selling its products in 150 countries through a network of approximately 4,900 independent dealers. New Holland had revenues of $5.7 billion in 1998, the company and its joint venture partners operate in 160 countries through a network of approximately 6,100 dealers and distributors.

Below: Ploughing with a four-wheel drive Massey-Ferguson 8170 tractor.

These pages: Loaders such as this Massey-Ferguson (**left**) are a relatively recent farming innovation and are typical of the trend to specialised farm machinery. Equally modern but more traditional is this PTO-driven hay mower (**below left**). Considerable weight has been added to the front axles of the two models of Massey-Ferguson seen ploughing (**right and below right**).

Above: The Massey-Ferguson 4200 series of tractors comprises eight models designed for the mixed and livestock farmer.

Left: The Massey-Ferguson 6120 is powered by an 86hp naturally aspirated Perkins diesel engine and features a 16-speed transmission.

Above Right: The Massey-Ferguson 4260 is powered by an in-line six-cylinder naturally aspirated Perkins diesel engine that produces 106hp and drives through a 12-speed transmission.

Right: The 4270 is a larger model in the same range. It is powered by a similar engine but produces 114hp .

The Massey-Ferguson 6100 series of tractors are designed for mixed farming and there are nine models in the range. These include the 77hp 6110 (**left**) and the 6140 (**below**) which are powered by different in-line four-cylinder Perkins diesel engines.

Above: The JCB Fastrac is made by the noted British manufacturer J. C. Bamforth and this one is seen making silage in a West Yorkshire, England field.

Europe

During the 1990s the developments that concerned European manufacturers of tractors and farming equipment were increasingly trans-Atlantic. In 1990, as has been already mentioned, the **Deutz-Allis** division was sold by its German owners to AGCO, who later reintroduced the brand as AGCO-Allis tractors. These machines are produced as one of their numerous brands of AGCO farming equipment.

FiatGeotech acquired Benati in 1991, which was later merged with Fiat-Hitachi. Fiat also acquired Ford New Holland Inc. and also merged it with FiatGeotech. The new company was named N. H. Geotech but became New Holland Inc in 1993.

Claas continued the manufacture of harvesting machines including those for specialised tasks such as sugar cane harvesting. The company claims that more than 80 crops are threshable with Claas combine harvesters, ranging from cereals to maize, rice, beans, sunflowers, grass and clover seed. Claas machines range from the 212hp (156kW) Model 35 to the 410hp (306kW) Model 95E. It also makes sugar cane harvesters and currently offers the cc 3000 and the Ventor, which are designed for harvesting burnt or green cane. Early in the history of mechanical harvesting of sugar cane Claas recognised the need to develop machines suitable for harvesting green cane. The current Claas range of balers for hay, silage and straw is wide and includes large square balers—known as Quadrant models—Rollant round balers, Variant variable chamber

461

Previous Page: This JCB Fastrac is seen with a converted semi-trailer moving large square bales from field to storage.

Left and Above: The JCB Fastrac—such as this 1135 model (**left**)—is capable of up to 50mph on the road; hence its name, an acronym for 'fast tractor'. JCB also manufacture handling and loading machinery including the Loadall (**above**).

balers and conventional Markant square balers. Self propelled foragers harvest grass or lucerne for drying, wilted grass silage and silage maize. The Claas Jaguar range is one of the large forager ranges suited to harvesting methods such as whole crop silage. Alongside the self-propelled models Claas also manufactures pull-type and mounted foragers, and green harvest machines. The Claas Saulgau plant is located in the heart of Europe's largest green harvest region.

Renault's 1990s tractor models included the Model 61RS, a low profile machine designed for operation in confined areas. It was powered by a three-cylinder diesel engine that produces 60hp, giving a maximum of 57hp at the PTO. The 80 TYX was a four-cylinder water cooled diesel powered tractor capable of producing 78hp. It was available as either a two-wheel or four-wheel drive model with 12 forward and reverse gears. The Model 106-

14 was a larger tractor with a six-cylinder diesel engine of 345cu in (5,656cc) displacement. It produced 96bhp and was available in both two- and four-wheel drive configurations.

Case-IH acquired a controlling interest—75 percent of the shares—in **Steyr** Landmaschinentechnik GmbH (SLT) from Steyr-Daimler-Puch AG. Steyr Landmaschinentechnik GmbH, was a tractor manufacturer based in St. Valentin, Austria, with annual revenues of $176 million. Another German company, the tractor maker **Fendt**, was acquired by the AGCO corporation in 1997. The Case-IH company has numerous plants around Europe; St. Valentin, Austria is where Steyr Tractors are made, while excavators, loaders and backhoes are produced in Crepy-En-Valois, France. Another French plant at Croix, produces agricultural and construction equipment cabs for Case while others at St. Dizier and Tracy-El-Mot produce transmissions

Left and Above: The 1980s and 1990s has seen ploughing become a more technologically advanced operation than it used to be. Contrast these massive four-wheel drive machines seen in a Scottish field with the grey Fergusons of the 1950s!

and hydraulic cylinders respectively. Tractors, gears and shafts are made in Doncaster, England while elsewhere in the same country mini-excavators and skid steer machines are made in Manchester and sprayers are made in Lincoln. In Neustadt, Germany forage harvesters, combines, square balers are produced.

Case-IH introduced a new mid-horsepower range of tractors, the MX series, but moved production of the MX from Neuss, Germany, to plants in Racine, Wisconsin, and Doncaster, England. Closure of the Neuss plant is a major step in Case's long-term restructuring program. Case also acquired Gem Sprayers Limited, a privately-owned maker of self-propelled and trailer-mounted

Above and Right: Ergonomically designed cabs, air conditioning and electronic systems have advanced farming techniques and increased productivity; Italian Lamborghini (**above**) and American John Deere (**right**) vehicles are pictured.

sprayers for agricultural applications. Gem was the leading supplier of sprayers in the United Kingdom with sales of $12 million. Case Credit and UFB Locabail SA formed a joint venture to provide financing for Case's European dealers and retail customers. The new venture is known as Case Credit Europe. Later Case announced plans to acquire a German company—Fortschritt—and the assets of two others. The acquisition is intended to give Case a broad range of conventional and rotary combines in Europe and significantly expand its line of harvesting equipment there.

The established British manufacturer **JCB** is noted for the production of a comprehensive range of machines including the famous 'digger' with both a front loader and back hoe. The company's range of machines during the 1990s included the Fastrac. The JCB Fastrac was a modern high speed tractor. It featured a unique all-round suspension system, a spacious ROPS and FOPS cab with air conditioning and passenger seat as standard. It had four equal-sized wheels. The Fastrac has a three-point implement mounting position

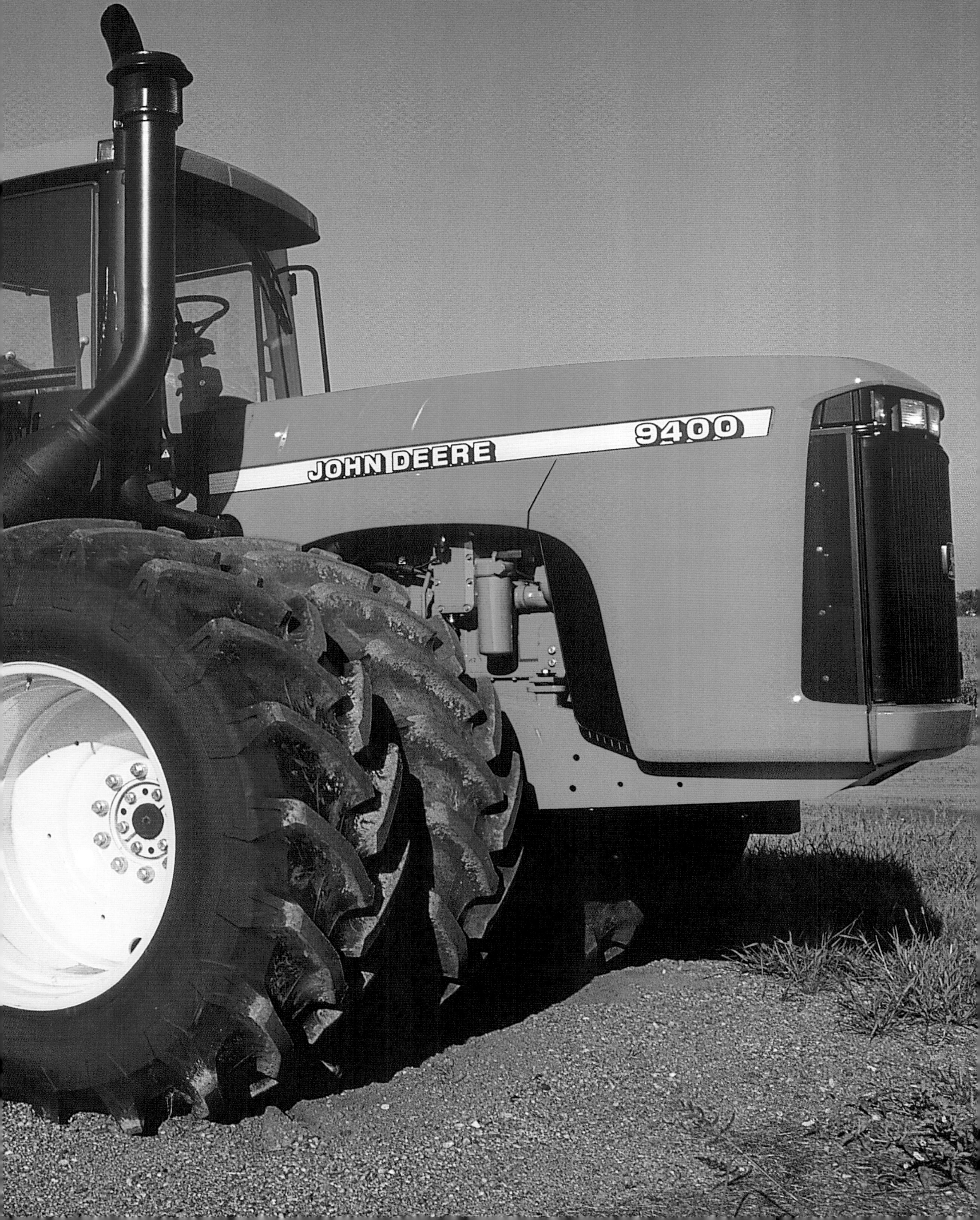
JOHN DEERE
9400

This Page: Italian tractor makers have prospered since World War II. Lamborghini (**main picture**) is now part of Same while Antares makes equally advanced tractors.

and the optional JCB Quadtronic four-wheel steering system is available on the 2115, 2125 and 2135 Fastrac models. This automatically switches between two-wheel steering and four-wheel steering for tight headland turns. Power comes from turbo diesel engines that produce 115hp–170hp (85.8kW–127kW). The diesel Perkins 1000 Series high torque engine used meets 'stage 1' off highway emission regulations, is quiet and fuel efficient. All models are turbocharged; the 2135 has a waste guard turbo, while the 2150 has a waste guard and intercooler. This is coupled to a three-speed Powershift transmission and an electronic transmission controller, which has been designed to ensure smooth changes under load. The suspension and chassis use equally advanced technology including a three-link front suspension that allows the tyres to tuck in against the chassis for a tight turning circle. Self-levelling on the rear suspension compensates for the additional weight of implements and self-levelling from side to side assists boom stability which is useful when working on hillsides. Optional adjustable wheel slip control, working in conjunction with a radar sensor that measures true ground speed, helps reduce tyre wear and soil smearing. Inside the cab is the JCB Electrical Monitoring System (EMS) which provides a performance assessment of the numerous machine functions on the dashboard. It includes the following: engine rpm, PTO rpm, selected Powershift ratio and a full range of warning indicators such as one indicating whether the front or rear PTO is selected. The axle drive shafts have double seals on the bearing cups to retain more grease for longer life. Improved axle drive shafts and cross serrated drive shaft location is designed to ensure durability and longer component life. The heavy duty front axles feature a solid bearing spacer between the pinion bearing for improved

Above and Inset: Landini is another successful Italian tractor maker. Currently there are more than 50 tractors in its range including the Discovery models and those made especially for vineyards.

Right: Same acquired the tractor-making portion of the Lamborghini company when it and the car maker were sold separately. This is a Premium 950 model.

durability. Soft engage differential-lock, fitted as standard to the rear axle, can be selected when wheels are spinning. It engages smoothly to protect the drive train and then automatically disengages when four wheel steering is operated or when the 'differential-lock cut-out' is selected.

Landini sold the controlling interest in its affairs to the Cameli Gerolimich Group. This move took Landini S.p.A. into the 1990s in the forefront of the tractor industry both in Italy and worldwide. Landini redesigned its tractors and offered a comprehensive range with the launch of the Trekker, Blizzard and Advantage series. As a result sales of Landini tractors exceeded 3,000 units in the company's export markets for the first time. In February 1994 Valerio and Pierangelo Morra, as representatives of the Argo S.p.A Family Holding Company, became President and Vice President of Landini S.p.A. respectively. They and Massey-

Left: Claas distributes the Caterpillar Challenger models of crawler tractor in Europe.

Below left: A Landini Discovery laden with grapes at work in an Italian vineyard.

Right: The latest Deutz-Fahr models display the contemporary curved styling of tractor body-work and ergonomically designed cabs.

Ferguson contributed to a substantial recapitalisation of Landini and in March 1994 Iseki joined Landini S.p.A. In December of the same year, Landini S.p.A. announced a net profit of seven billion Lira and an increase of tractor sales by more than 30 percent over the year previous. Change continued into the following year when, in January 1995, Landini acquired Valpadana S.p.A., a noted brand in the Italian agricultural machinery sector. During the following month the company made an agreement for the distribution of the Landini products in North America through the AGCO Corporation sales network and the renewal of the agreement to be the sole supplier of specialised wheeled and crawler tractors to AGCO.

Other overseas markets were not ignored, and in March 1995 Landini Sur America was opened in Valencia, Venezuela. This company was aimed at promoting the Landini brand in Latin America, a market that was considered to have potential for increased sales of tractors. In 1995 Landini sales, including those of Massey-Ferguson imported tractors and the Valpadana tractors manufactured in the San Martino plant, reached a total of 14,057 units, of which 9,415 machines were sold under the Landini name. During 1996 in order to meet the growing demand for its tractors a new assembly-line was installed in the Fabbrico, Italy factory. The new line was intended to double the factory's production capacity. In San Martino in 1996 a new factory was opened; it was entirely dedicated to machining operations, gear manufacture and prototype component assembly. The current Landini range includes 50 different models, with power outputs ranging from 43 to 123 PTOhp. Landini tractors are currently distributed in the USA by AGCO and the current range includes the following models: Advantage, V, F and GT series, HC

Far left: Fendt Favorit and Farmer models are still in production albeit considerably upgraded and under the auspices of the AGCO group.

Above: The curved styling seen on this Deutz-Fahr Agcotion 100 increases visibility from the cab.

Left: Four-wheel drive traction is enhanced by the provision of additional weights forward of the front axle.

series, Trekker series, 60 series, 80 series, Blizzard, Legend and Globus series.

Belarus Tractors are made in the former USSR state of Belarus and have exported tractors widely. Belarus is now a member of the CIS—Commonwealth of Independent States—and offers numerous tractors in a variety of configurations. The current Belarus range of four-wheel drive tractors includes the Model 1221, that produces 130hp, the 115hp 1025, the 105hp 952 and the 862 that produces 90hp. The two-wheel drive line of Belarus tractors ranges from the 105hp 950 machine to the 90hp 900 model.

The **Minsk Tractor Works** distributes its tractors through agricultural machine distributors in 35 countries around the world. 1990s production included models such as the MTZ-320, MTZ-682, MTZ-1221 and MTZ-920.

Zetor is a Czechoslovakian company but versions of its tractors are assembled in numerous countries around the world including Argentina, Burma, India, Iraq, Uruguay and Zaire. Zetor tractors are marketed and currently sold worldwide through two major distribution channels. The first of these is companies of the Motokov Group that has offices around the world while the second is the John Deere company dealer network. This was made possible through agreements reached between the Motokov Group, John Deere and Zetor. In 1993, under a distribution agreement reached in this year with John Deere, Zetor was able to distribute a lower-priced line of 40hp to 85hp tractors in what are generally considered to be emerging markets, starting with selected areas of Latin America and Asia. In 1997 the Zetor product line included ten different tractors. These ranged from the 46hp Models 3320 and 3340 to the 90hp Models 9640 and 10540.

Left and Below: European tractor production is as advanced as that of the United States although tractor makers such as Deutz-Fahr (**left**) can now be considered as multinational. Valmet, also European, makes specialised forestry tractors (**below**).

Previous Page: Valmet is a Finnish tractor maker. This is the current four-wheel drive model of tractor.

Above and Right: Valmet of Finland makes farm tractors such as these models (**top and above**) while another former Eastern Bloc tractor maker is Zetor from Czechoslovakia. The 7245 is one of the latter's more recent tractors (**right and above right**).

Next Page: Ursus is a long established Polish tractor maker. This recent 3512 model is among the company's exports.

This Page: Ursus tractors are diesel-powered (engine detail **inset**) while the standardisation of implements in the tractor industry is demonstrated by the fact that this Ursus 4514 has been fitted with a Massey-Ferguson hydraulic front loader.

Left and Above: Kubota is an established Japanese tractor maker that specialises in compact machines which are ideally suited to grounds and lawn maintenance (**above**), and nursery use (**left**), although numerous implements including front loaders and back hoes are available (**top**).

Right: The Kubota B1550 (**right**) was a compact tractor ideal for grounds maintenance. It features a small scale three-point linage and PTO shaft (**below right**).

Opposite Page: Tractor pulling is a sport that has a worldwide following. This multi-engined machine is based in Holland and competes in European events.

Rest of the World

Iseki is a Japanese manufacturer of tractors. It makes both compact models and larger conventional machines. In the former category are tractors such as the TX 2140 and 2160 models. They are powered by three-cylinder water-cooled engines of 776 and 849cc (47.3 and 51.8cu in) respectively in both two- and four-wheel drive forms, and are suited for use with a range of implements. One of Iseki's larger machines is the Model T6500, which is powered by a 3,595cc four-cylinder water-cooled diesel engine. The transmission offers 20 forward and five reverse gears. It has equal-sized wheels all round to enhance ground clearance and make it suitable for use with crops where high clearance is required.

Kubota is another Japanese tractor maker. One 1990s' Kubota product was the compact B7100DP, a three-cylinder powered tractor that displaces 46.5cu in (762cc) and produces 16hp. It also featured four-wheel drive, independent rear brakes and a three-speed PTO. In the current range are models including the 656cc (40cu in) G1700 and 719cc (43.9cu in) G1900-S machines with four-wheel steering. The company also offers the T1560 and TG1860 models of 423 and 719cc (25.8 and 43.9cu in) displacement respectively. All are diminutive machines of the ride-on mower type. The largest Kubota compact tractor is the Grandel L series of L3300, L3600 and L4200 models. Of these, the L4200 is the largest with an

engine displacement of 2,197cc (134cu in) that produces 45.3hp; it is intended for service applications.

Mahindra continued to make tractors in India while tractors badged as '**American Harvester**' machines are currently made under contract in the People's Republic of China and imported into the United States by Farm Systems International. The

engines used in American Harvester machines incorporate features such as removable cylinder sleeves and forged pistons aimed at ensuring the longevity of the tractors. The Model 504 is one of a range that is designed to work for 6,000–11,000 hours between overhauls and is also designed to be sufficiently powerful but compact and fuel-efficient. The 504 is fitted with a low pollution diesel engine coupled to an eight-speed transmission with four-wheel drive capability. Options include a two-wheel drive transmission with an adjustable width, row-crop front end. The 504 has a full size, mechanical PTO. Also standard is a built-in, hydraulic PTO and a full size, three-point hydraulic hitch. The range of models is available with engines that produce from 18hp to 50hp. All have a 12V electrical system with emergency hand crank starting, glow plugs and compression release for cold weather. There is a built-in auxiliary hand pump to purge fuel lines. The model 504 is 141.34in long, 65.35in wide and the height to the steering wheel is 56.4in. The wheelbase is 80.7in while ground clearance is 14.2in.

The American Harvester Model 250 is of smaller overall dimensions with a wheel base of 60in. It is primarily intended for use implements including rotary cultivators, rotary disc ploughs, rotary harrows and reaping-machines as well as ordinary ploughs and harrows. To enable them to do this the 250s have a system of 'live hydraulics' and large diameter tyres to aid traction in fields. The tractors can also be used as the power for irrigation and drainage equipment, threshing machines and rice mills or to drive trailers. The Model 250 is

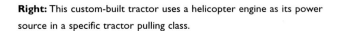
Right: This custom-built tractor uses a helicopter engine as its power source in a specific tractor pulling class.

powered by a vertical, water-cooled, four-stroke, three-cylinder diesel. Its displacement is 1,432cc (87.4cu in) achieved through a bore and stroke of 3.4x3.8in. The compression ratio is 22:1 and the engine produces 25hp at 2,500rpm and a maximum torque of 80.8lb/ft. The transmission has eight forward and two reverse gears and is assembled with a dry, single-plate clutch. The gearbox is fitted with a differential lock to assist traction in wet or heavy soil. The drum brake is of the internal expanding shoe type while steering is of a traditional peg and worm design. The Model 250 has 6.0x15 and 8.3x24 tyres, front and rear respectively and its hydraulic linkage has a maximum lift capacity of 4,140lb.

During this decade **Case** acquired Austoft Holdings Limited, the world's largest manufacturer of sugar-cane harvesting equipment based in Bundaberg, Australia. With annual sales of $74 million, Austoft considerably strengthened Case's business in Australia as well as its presence in key emerging agricultural markets. Uzbekistan purchased approximately $80 million of Case-IH agricultural equipment, the company's largest sale to date to this growing Central Asian agricultural equipment market. To further develop business in Central Asia, Case formed a joint venture in Uzbekistan to manufacture two-row cotton-pickers. The venture increases the availability of Case's advanced cotton-pickers to Uzbekistan, a major producer and exporter of cotton. Case holds a majority share in the UzCaseMash venture, based in the capital city of Tashkent. During 1996 **John Deere** also made its the largest single agricultural sale ever. This was an order worth $187 million for combine harvesters to be supplied to the Ukraine.

Left: A John Deere tractor puller in the US.

Above Right: A Dutch tractor puller powered by no fewer than three large displacement V8 engines.

Right: Another Dutch tractor puller. There are numerous classes for horsepower and engine types.

Above and Right: Tractor pulling in California.

Opposite Page, Above: Four-wheel drive Case tractor .

Opposite Page, Below: Fermec machines in California.

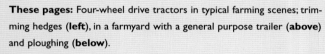

These pages: Four-wheel drive tractors in typical farming scenes; trimming hedges (**left**), in a farmyard with a general purpose trailer (**above**) and ploughing (**below**).

Above: Ploughing in an English field.

Far right, inset: A New Holland tractor in an English farmyard.

Right: A John Deere 4700 crop sprayer, note high crop clearance under axles.

Previous page: John Deere 6400 with a dual-directional three-furrow plough.

Next page: A four-wheel drive John Deere 2130 equipped for crop spraying.

These pages: John Deeres at work; in a West Yorkshire, England field (**above**), and trimming a hedge alongside a road in England (**right**) note the arrow to warn approaching traffic.

These pages: Tractors at work—with a towed implement (**top**), with a hydraulic front loader stacking round bales (**above**) and with a towed New Holland baler (**right**).

Above and Left: Renault, like other makers, makes tractors suitable for numerous applications including a rear pallet lift (**above**), for operating towed implements such as this harrow (**left**) and this Claas square baler (**below left**).

Right: Four-wheel drive tractors with a hydraulic front loader (**above**) and a seed planter (**below**).

INDEX